# Automated Theorem Proving

**Springer**
*New York*
*Berlin*
*Heidelberg*
*Barcelona*
*Hong Kong*
*London*
*Milan*
*Paris*
*Singapore*
*Tokyo*

Monty Newborn

# Automated Theorem Proving

## Theory and Practice

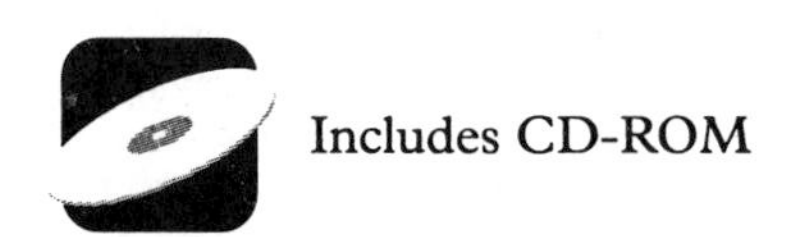
Includes CD-ROM

Springer

Monty Newborn
School of Computer Science
McGill University
3480 University Street
Montreal, Quebec H3A 2A7
Canada
newborn@cs.mcgill.ca

With 7 illustrations.

Library of Congress Cataloging-in-Publication Data
Newborn, Monty.
Automated theorem proving : theory and practice / Monty Newborn.
p. cm.
Includes bibliographical references and index.
ISBN 0-387-95075-3 (alk. paper)
1. Automatic theorem proving. I. Title.
QA76.9.A96 N49 2000
004'.01'5113—dc21 00-056315

Printed on acid-free paper.



Production managed by MaryAnn Brickner; manufacturing supervised by Jeffrey Taub.
Photocomposed copy prepared from the author's Adobe Pagemaker and Microsoft Excel files.
Printed and bound by Maple-Vail Book Manufacturing Group, York, PA.
Printed in the United States of America.

9 8 7 6 5 4 3 2 1

ISBN 0-387-95075-3 SPIN 10750623

Springer-Verlag New York Berlin Heidelberg
*A member of BertelsmannSpringer Science+Business Media GmbH*

# Preface

As the 21st century begins, the power of our magical new tool and partner, the computer, is increasing at an astonishing rate. Computers that perform billions of operations per second are now commonplace. Multiprocessors with thousands of little computers — relatively little! — can now carry out parallel computations and solve problems in seconds that only a few years ago took days or months. Chess-playing programs are on an even footing with the world's best players. IBM's Deep Blue defeated world champion Garry Kasparov in a match several years ago. Increasingly computers are expected to be more intelligent, to reason, to be able to draw conclusions from given facts, or abstractly, to prove theorems — the subject of this book.

Specifically, this book is about two theorem-proving programs, THEO and HERBY. The first four chapters contain introductory material about automated theorem proving and the two programs. This includes material on the language used to express theorems, predicate calculus, and the rules of inference. This also includes a description of a third program included with this package, called COMPILE. As described in Chapter 3, COMPILE transforms predicate calculus expressions into clause form as required by HERBY and THEO. Chapter 5 presents the theoretical foundations of semantic tree theorem proving as performed by HERBY. Chapter 6 presents the theoretical foundations of resolution–refutation theorem proving as performed by THEO. Chapters 7 and 8 describe HERBY and how to use it. Chapters 9 and 10 parallel Chapters 7 and 8, but for THEO. Chapter 11 and 12 discuss the source code for the two programs. The final chapter, Chapter 13, briefly examines two other automated theorem-proving programs, Gandalf and Otter.

In the 1970s and 1980s, the author was involved in the design of chess programs. His program OSTRICH competed in five world computer chess championships dating back to 1974, when it narrowly missed defeating the Soviet program KAISSA in the final round of the first World Computer Chess Championship in Stockholm. Many of the lessons of programming chesscarry over to the field of automated theorem proving. In chess, a program searches a large tree of move sequences looking for the best line of play. In theorem proving, a program also searches a large tree of inferences — rather than moves — looking for that special sequence that yields a proof.

At the heart of both problems is the exponential nature of the search tree and the use of various algorithms and heuristics that direct the search toward the more relevant parts of the tree. Chess programs use various algorithms to narrow the search space such as the alpha-beta algorithm, windowing algorithms, algorithms that take advantage of move transpositions, and iteratively deepening depth-first search. Some programs also use various heuristics, best defined as rules of thumb, to further narrow the search, although experience has shown that one must use extreme caution in this case. The heuristics in early chess programs were far too unreliable, and as the former World Champion Mikhail Botvinnik once said, often "threw away the baby with the bath water." Most heuristics are dangerous because as the level of play goes up, the number of exceptions to any heuristic increases as well. "Beautiful" moves often violate the relatively simpleminded heuristics used in programs.

The search techniques used to prove theorems are very similar to those used in chess programs. The theorem-proving program THEO contained in this package uses iteratively deepening depth-first search, hash tables for reducing the search space, and other algorithms to narrow the search space without sacrificing the ability to find a proof if one exists. Several heuristics are also normally used which do sacrifice this ability, although the user can choose not to use them. The second theorem-proving program HERBY constructs large semantic trees in its effort to prove a theorem using various heuristics to guide the process.

These two programs are meant to familiarize the reader with search techniques used in theorem-proving programs, to permit experiments with two capable theorem-proving programs, and to provide the source code so that the reader can attempt to improve it. In 1989, THEO, then called THE GREAT THEOREM PROVER, or TGTP, first appeared, and over the years it was used as a text/program at several universities and research centers. The latest version, recently renamed THEO, contains a far stronger theorem-proving program, a more extensive text to go along with it, and moreover, this time source code is available. As a theorem-proving program, THEO is quite sophisticated. HERBY is less capable as a theorem-proving program, but its approach is particularly simple and seems to have considerable potential. Both programs have participated in the Conference on Automated Deduction's competitions for such programs, as discussed in Chapter 13.

The package, consisting of software and text, can serve as instructional material for a course on theorem proving at either the undergraduate or graduate level. It can also serve as supplemental material for an introductory course on artificial intelligence. The package includes almost two hundred theorems for the student. Some are very easy and others are very difficult. There are many examples scattered throughout the text, and there are exercises at the end of every chapter. In addition to the theorems included, several thousand theorems that are used by the automated theorem-proving research community can be obtained by ftp, as is explained in Chapter 1.

The source code provided in this package has evolved over a ten-year period. Every effort has been made to make it readable. Every file lists the functions contained, and every function has a header that lists who calls it, who it calls, its arguments, and what it returns. Students should be able to modify the code as a class project.

The author would like to thank a number of people who have helped to develop THEO and HERBY. In particular, former McGill University students Paul Labutte, Patrice Lapierre, and Mohammed Almulla and current students Choon Kyu Kim and Paul Haroun deserve thanks. In addition, countless McGill students that the author has had in his classes must be thanked for their many suggestion on how to improve the programs. The author had many problems over the years in developing the software; thanks for help is extended to the System Support Group of the School of Computer Science at McGill. Lastly, the author's two daughters, Amy and Molly, must be given a special thanks for tolerating the passion of their father for this esoteric subject.

In the weeks leading up to the publication of this book, the author discussed the manuscript with Gabby Silberman, head of IBM Toronto's Center for Advanced Studies. He offered to add IBM's C compiler for AIX, Version 4.4, to the CD-ROM. IBM's Joe Wigglesworth checked out the theorem-proving software and confirmed that it was compatible with the IBM software. Springer-Verlag and IBM, as you can see, worked out an agreement whereby the compiler is included on the CD-ROM. The author would like to thank IBM and Springer-Verlag for this.

The author wishes the reader many hours of interesting learning and experimentation. Any suggestions for improving this package for the next version can be sent to the author via e-mail at newborn@cs.mcgill.ca.

Montreal, Canada
September 2000

Monty Newborn

PROOF
THEO

# Contents

P0: a is an animal
P1: a is a wolf
P2: a is a fox
P3: a is a bird
P4: a is a caterpillar
(Ax)(P1x → P0x) & (Ex) P1x
(Ax)(P2x → P0x) & (Ex)
P5: a is a snail
Q0: a is a plant
Q1: a is a grain

# 1 A Brief Introduction to COMPILE, HERBY, and THEO

This book is about how computers prove theorems. In particular, it is about how two programs, HERBY and THEO, carry out this fascinating process. This first chapter briefly introduces the reader to these two programs and to a third called COMPILE. COMPILE prepares a theorem, transforming it into the required format, for HERBY and THEO. Once in the required format, HERBY and THEO can then attempt to find a proof. HERBY is a semantic-tree theorem-proving program. THEO is a resolution–refutation theorem-proving program. These three programs are written in ANSI C and compile using standard C compilers. They run under UNIX (and Linux, Solaris, FreeBSD, and AIX). IBM's C compiler for AIX, Version 4.4, is included on the CD-ROM accompanying this text.

The theorems of interest must be expressed in the language of predicate calculus, a mathematical language commonly used by scientists to express facts precisely. A statement in this language is called a well-formed formula. A theorem consists of a set of well-formed formulas, some of which are axioms—"givens"—true statements about the theorem's domain of discourse; others constitute the "hypotheses" and the "conclusion."

The hypotheses and the conclusion are sometimes expressed as an "if–then" statement, where the "if" side can be considered as a given, and included among the axioms. The conclusion—that is, what we wish to prove—is the "then" side. For example, one theorem in Euclidean geometry states that: if two triangles ABC and abc have two equal sides, say side AB is equal to side ab and side BC is equal to side bc, and if the included angles ABC and abc are equal, then the two triangles ABC and abc are identical. In this example, the "if" side is the fact that two sides and the included angle of triangle ABC are equal to two sides and an included angle of triangle abc. The "then" side is the conclusion that the two triangles are identical .

process and crucial to the success of a theorem-proving program. If too few axioms are chosen, the one necessary for a proof might be omitted and a proof never found. If too many are chosen, the program might take too much time considering irrelevant axioms and never find a proof.

In Chapter 2, the problem of creating well-formed formulas from English statements is considered. A number of examples are presented. If you have a good mathematics background, you might like to take a good look at the axioms used to prove theorems in group theory in Section 2.5. In Section 2.6, the axioms used to prove theorems in Euclidean geometry are presented.

## 1.1 COMPILE

Once the axioms and conclusion have been expressed as well-formed formulas, they must next be translated to clauses by COMPILE in order to be compatible with HERBY and THEO. COMPILE first negates the conclusion and then translates the axioms and negated conclusion to clauses. The user can bypass COMPILE by creating a file of the axioms and negated conclusion as clauses, and this is often done. Chapter 3 describes COMPILE.

### 1.1.1 Creating an executable version of COMPILE

The source files for COMPILE are in the directory COMPSC. There are eight -.c files, one -.h file, and a makefile. To create an executable version of COMPILE, type the command:

```
make
```

After several seconds, the executable file, COMPILE, will be ready.

### 1.1.2 Running COMPILE

To run COMPILE, type:

```
compile
```

COMPILE then responds with:

\What file do you want to compile?

You should then enter the name of the file containing the well-formed formulas that you wish to compile. COMPILE then translates the wffs to clauses, printing the results on the screen and saving them in a file on disk.

## 1.2 HERBY

The theoretical foundations for semantic-tree theorem proving are provided in Chapters 2 through 5. Chapters 2 through 4 provide the theoretical foundations for both HERBY and THEO. Chapter 5 is specifically devoted to HERBY. It describes how to construct a semantic tree when trying to prove a theorem. If the tree is closed, or, in other words, has finite size, the theorem is proved. It is pointed out that good heuristics for ordering the selection of what are called atoms are needed to carry out this procedure effectively. HERBY is a semantic-tree theorem-proving program that makes this attempt. HERBY's heuristics form the material in Chapter 7. Chapter 8 describes how to use HERBY. Chapter 11 describes the source code for HERBY.

### 1.2.1 Creating an executable version of HERBY

The source files for HERBY are in the directory HERBYSC. There are 11 -.c files, one -.h file, and a makefile. To create an executable version of HERBY, simply type the command:

```
make
```

After several seconds, the executable file, HERBY, will be ready.

### 1.2.2 Running HERBY

To run HERBY using default options, type:

```
herby
```

HERBY then responds with:

**Enter name of theorem (Type '?' for help):**

You should then enter the name of the file containing the theorem you want to prove. HERBY then sets out to attempt to construct a closed semantic tree, thus establishing a proof of the theorem.

## 1.3 THEO

As stated previously, Chapters 2 through 4 and Chapter 6 provide the theoretical foundations for THEO. Chapter 6 describes resolution–refutation theorem proving and proves that every theorem has a resolution–refutation proof. It also proves that every theorem has a linear resolution–refutation proof. It is this type of proof that THEO searches for. The search procedure carried out by THEO is described in Chapter 9, and Chapter 10 describes how to use the program. Chapter 12 considers the source code for THEO.

### 1.3.1 Creating an executable version of THEO

The source files for THEO are in the directory THEOSC. There are 12 -.c files, two -.h files, and a makefile. To create an executable version, THEO, simply type the command:

make

After approximately one minute, the executable file, THEO, will be ready.

### 1.3.2 Running THEO

To run THEO using default options, type:

theo

THEO then responds with:

**Enter name of theorem (Type '?' for help):**

You should then enter the name of the file containing the theorem you want to prove. THEO then sets out to attempt to find a resolution–refutation proof of the theorem.

## 1.4 The Accompanying Software

The software accompanying this text is contained on a CD-ROM in a directory called TP, which, in turn, contains the seven directories described here. You should begin by transferring them to your root directory as shown in Figure 1.1. The seven directories are:

COMPSC is a directory containing source files for COMPILE. There are eight -.c files and one -.h file. A makefile is included.

HERBYSC is a directory containing source files for HERBY. There are 11 -.c files and two -.h files. A makefile is included.

THEOSC is a directory containing source files for THEO. There are 12 -.c files and two -.h files. A makefile is included.

WFFS is a directory containing 21 files of well-formed formulas and theorems expressed as well-formed formulas.

THEOREMS is a directory of Stickel's test set of 84 theorems used by researchers to test theorem-proving programs.

GEOMETRY is a directory of Quaife's 66 theorems on Euclidean geometry.

THMSMISC is a directory of 32 miscellaneous theorems.

The well-formed formulas and theorems contained in the directories WFFS, THEOREMS, GEOMETRY, and THMSMISC are listed in Appendix B.

| | |
|---|---|
| root:/TP/WFFS | 21 miscellaneous wffs |
| THEOREMS | 84 theorems from Stickel's test set |
| GEOMETRY | Quaife's 66 theorems on Euclidean geometry |
| THMSMISC | 32 miscellaneous theorems |
| COMPSC | Source files and makefile for COMPILE |
| THEOSC | Source files and makefile for THEO |
| HERBYSC | Source files and makefile for HERBY |

Figure 1.1. The TP directory.

In addition to the software provided, you are encouraged to acquire the TPTP (Thousands of Problems for Theorem Provers) Problem Library. This can be obtained by anonymous ftp from the Department of Computer Science, James Cook University, at coral.cs.jcu.edu.au.

## Exercises for Chapter 1

1.1. Download the TPTP Problem Library to your system. Do this as follows:

(a) Obtain the TPTP Problem Library

```
> mkdir tptp
> cd tptp
> ftp –i ftp.cs.jcu.edu.au
    When it requests your name, type ftp
    When it requests your password, type your email address
> cd pub/research/tptp–library
> bin
> mget *
> quit
```

(b) Install the TPTP Problem Library

```
> gunzip –c TPTP–v2.3.0.tar.gz | tar xvf –
> TPTP–v.2.3.0/Scripts/tptp1T_install
    <... then answer questions that are asked>
```

Notes: (1) TPTP–v2.3.0.tar.gz is a large file, approximately 20 megabytes. (2) You may need to edit the file tptp1T_install as follows: change the instruction: "set Perl=perl" to "set Perl=perl5" if your computer uses this newer version.

1.2. How many different categories of theorems are there in the TPTP Problem Library? What do the abbreviations for each of the categories stand for?

1.3. How many theorems are there in the geometry (GEO) directory of the TPTP Problem Library? Examine GEO004–1.p. (Note: when you examine GEO004–1.p, you will find that most of the clauses are in the "include" files, EQU001–0.ax, GEO001–0.ax, and GEO001–0.eq.) How many clauses are axioms? How many are hypotheses? How many constitute the negated conclusion? What the names of the predicates? What are the names of the functions? In words, what are the hypothesis and negated conclusion (conjecture, in the TPTP Problem Library's vernacular)?

# 2 Predicate Calculus, Well-Formed Formulas, and Theorems

When proving a theorem, it is first necessary to write the axioms, the hypotheses, and the conclusion. Deciding what axioms to choose in the first place is crucial to the success of a theorem prover, but that problem is peripheral to the presentation that follows. In general, there is no procedure for deciding what axioms are necessary or sufficient. In some problem domains, standard sets of axioms are known and used. For example, in group theory and in Euclidean geometry, many researchers use the axioms given in Sections 2.5 and 2.6, respectively.

The language used by many theorem-proving programs is first-order predicate calculus or, more simply and henceforth, predicate calculus. This chapter describes how the user can write statements in this language. The syntax of predicate calculus statements is presented in Section 2.1. Examples of predicate calculus statements are given in Section 2.2 along with their English equivalents. The process of converting statements in English to statements in predicate calculus is considered in Section 2.3. Section 2.4 defines a number of important terms. Section 2.5 presents a set of axioms that can be used to prove theorems in group theory. Section 2.6 presents a set of axioms that can be used to prove theorems in Euclidean geometry.

## 2.1 The Syntax of Well-Formed Formulas

In first-order predicate calculus, a statement is called a well-formed formula (wff). A wff is interpreted as making a statement about some domain of discourse. The syntax of wffs consists of logical operators, quantifiers, punctuation marks, terms, predicates, and literals defined here as follows:

**Logical operators**: **&** (and), **|** (or), **~** (negation), **=>** (implication), **<=>** (if and only if).

**Quantifiers**: @ (universal quantifier), ! (existential quantifier). The more conventional symbols, ∀ and ∃, are not used because they are not available on standard computer keyboards.

**Punctuation marks**: **,** (comma), **(** (left parenthesis), **)** (right parenthesis), **{** (left bracket), **}** (right bracket), **:** (colon), and **.** (period).

**Term:** A term is a variable, a constant, or a function with arguments that are themselves terms. Variable, constant, and function are defined next.

**Variable**: A variable represents potentially any element from the domain of discourse. It is represented by a string of letters and digits and underscores beginning with a letter. It is good practice to begin the name of a variable with a letter near the end of the alphabet, such as x, y, or z but there is no rule requiring this. Variables are distinguished from constants by the context in which they appear. This point is clarified in the last paragraph of this section.

**Constant**: A constant is a specific element from the domain of discourse. It is represented by a string of letters and digits and underscores beginning with a letter. Examples of constants are: a, b, c, molly, and amy. It is good practice to begin the name of a constant with a letter near the beginning of the alphabet, such as a, b, or c, but there is no rule requiring this.

**Function:** A function maps elements in the domain of discourse to other elements in the domain. It is represented by a string of letters and digits and underscores beginning with a letter. A function has one or more arguments. Parentheses enclose the arguments of the function, and commas separate individual arguments. The arguments are terms. Examples of specific functions are: abs(x), the absolute value function; smaller(x,y), a function that maps x and y to the smaller of the two; and min(x,y,z), a function that maps x, y, and z to the smallest of the three.

**Predicate:** A predicate is a relation on the domain of discourse. The relation is either TRUE or FALSE within the domain. A predicate is represented by a string of letters and digits and underscores beginning with a letter. A predicate has zero or more arguments. The arguments are terms. Parentheses enclose the arguments, and commas separate the arguments. Examples of predicates are: above(a,b) (read "a is above b"), animal(child_of((Jerry)) (read "the child of Jerry is an animal"), larger_than(square(x),x) (read "the square of x is larger than x") and hot (read "it is hot", a predicate with no arguments).

**Literal:** A literal is a predicate or the negation of a predicate.

**Well-formed formula:** A well-formed formula (wff) is defined recursively as follows:

A literal is a wff.
If w is a wff, then so is the negation of w, ~w.
If w and v are wffs, then so are: w | v, w & v, w => v, w <=> v.
If w is a wff, then, for any variable x, so are: @x: w, !x: w.

The quantifiers, @x and !x, are said to have **scope** over w, and x is called a **quantified variable.** Symbols representing variables can be distinguished from those representing constants: symbols representing the former are always quantified, whereas those representing the latter never are. Operator precedence for wffs is as follows (from low to high): <=>, =>, |, &, ~. Thus, for example, P | ~Q => R is equivalent to {P | {~Q}} => R.

## 2.2 Examples of Well-Formed Formulas

Seven wffs and their English equivalents are given in Figure 2.1.

## 2.3 Creating Well-Formed Formulas from Statements in English

The process of creating wffs from statements in English is similar to that of translating from English to any other language. Although there are some general rules, there is no algorithm for carrying out this process. When creating wffs from statements in English, it is best to begin by identifying and giving meanings to predicates, functions, ,variables, and constants. Three examples are presented in Figure 2.2.

## 2.4 Interpretations of Well-Formed Formulas

Consider the wff @x: G(sq(x),x). This wff can be interpreted in an infinite number of ways. Here are three: (1) Let @x refer to the domain of positive integers, sq denote the "square" function (i.e., sq(3) = 9, etc.), and G denote the "greater than" predicate. (2) Same as (1) except that the domain of discourse excludes +1. (3) Same as (1) except that "sq" is interpreted as the "square root" function.

Example 1: Cat(Floyd) reads "Floyd is a cat."

Example 2: Cat(Floyd) => Smart(Floyd) reads "If Floyd is a cat, then Floyd is smart." Note that this is equivalent to saying that "Floyd is not a cat or Floyd is smart," (i.e., ~Cat(Floyd) | Smart(Floyd)) or "Floyd is smart or Floyd is not a cat," or even "If Floyd is not smart, then Floyd is not a cat," (i.e., ~Smart(Floyd) => ~Cat(Floyd)).

Example 3: Cat(Floyd) | ~Cat(Floyd) reads "Floyd is a cat or Floyd is not a cat." This wff is a tautology. More later about tautologies.

Example 4: @x: { Cat(x) | Dog(x) } & Lives_on(x,Easy_Street) => Happy(x) reads "For all x, if x is a cat or dog and x lives on Easy Street, then x is happy," or less literally, "All cats and dogs living on Easy Street are happy."

Example 5: @x@y@z: Equal(x,y) & Equal(y,z) => Equal(x,z) reads "For all x, y, and z, if x equals y and y equals z, then x equals z."

Example 6: @x!y: ~Prime(x) => Prime(y) & Divides(y,x) & Less_than(y,x) reads "For all x, there exists a y such that if x is not a prime, then y is prime and y divides x and y is less than x."

Example 7: @x@y: set(x) & set(y) => {Equal(x,y) <=> {@u: in(u,x) <=> in(u,y)}} reads not quite as literally as the previous examples: "Two sets x and y are equal if and only if their members are the same."

Figure 2.1. Examples of wffs and their English equivalents.

An interpretation of a wff **satisfies** the wff if the wff has a logical value of TRUE under that interpretation. Of the three interpretations given earlier for @x: G(sq(x),x), only the second satisfies the wff. The first **fails to satisfy** the wff (or simply **fails**) because the sq(1) is not greater than 1. The third fails because the square root of x is not greater than x for x greater than 1. A wff is **satisfiable** if some interpretation satisfies it. It is said to be consistent with that interpretation.

If all interpretations satisfy a wff, the wff is called a **tautology**. For example, @x@y: E(x,y) | ~E(x,y) is a tautology; so is @x@y: E(x,y) | ~E(x,y) | E(sq(x),sq(x)).

If an interpretation given to a set of wffs makes each wff have a logical value of TRUE, then that interpretation **satisfies** the set of wffs. A wff W is said to **logically follow** from a set S of wffs if every interpretation satisfying

S also satisfies W. For example, @x: Furry(x) logically follows from the set of wffs {@x: Furry(x) | ~Dog(x), @x: Dog(x)}, as does Likes(Floyd,Linda) from the set consisting of the single wff {@x@y: Likes(Floyd,y) | ~Likes(x,Linda)}.

If some wff W logically follows from some set S of wffs, and S contains wffs that are tautologies T1, T2, ..., Tn, then W also logically follows from the smaller set of wffs S – {T1, T2, ..., Tn}. Thus, when trying to show that W logically follows from S, wffs that are tautologies can be eliminated from S, simplifying the problem of showing that W logically follows from S.

Finally, a **theorem** is a set of wffs; some wffs are "givens" and others, often only one, are "to prove." The givens are called **axioms**. The "to prove" usually involves **hypotheses** and a **conclusion** expressed in the form of an

Example 1: Translate to a wff: "All men except baseball players like umpires."
Solution: Choose man(x), baseballplayer(x), umpire(x), and likes(x,y) as predicates and give them the following interpretations: x is a man, x is a baseball player, x is an umpire, and x likes y. Then:

```
@x@y: man(x) & ~baseballplayer(x) & umpire(y) => likes(x,y)
```

or equivalently:

```
@x: man(x) & ~baseballplayer(x) => {@y: umpire(y) => likes(x,y)}
```

Example 2: Translate to a wff: "If x and y are nonnegative integers and x is greater than y, then $x^2$ is greater than $y^2$."
Solution: Choose nonnegint(x), greaterthan(x,y) as predicates and sq(x) as a function.

```
@x@y: nonnegint(x) & nonnegint(y) & greaterthan(x,y) =>
                                  greaterthan(sq(x),sq(y))
```

Example 3: Translate the following statement from set theory to a wff: "For all x, y, w and z, if z is in the intersection of x and y, and w is a member of z, then w is a member of x and w is a member of y."
Solution: Choose intersection(x,y,z) to denote that z is in the intersection of x and y, and member(x,y) to denote that x is a member of y.

```
@x@y@w@z: intersection(x,y,z) & member(w,z) =>
                          member(w,x) & member(w,y)
```

Figure 2.2. Examples of the creation of wffs.

"if this is true, then this follows" statement. The "if" part is assumed to be true, and the "then" part is what must be shown to be true as well.

Let the wffs that are the axioms and the hypotheses be denoted by S and the wffs that are the conclusion be denoted by W. If W logically follows from S, then there is no interpretation that satisfies both S and ~W, the **negated conclusion**; that is, S and ~W are **unsatisfiable.** Thus, it is possible to prove that W logically follows from S if it can be shown that S and ~W are unsatisfiable. This is, by definition, a **proof by contradiction** of W given S and is the proof technique used by THEO and HERBY. For example, it is possible to show that W = Likes(Paul,Tracy) logically follows from S = {@x: Girl(x) => Likes(Paul,x), Girl(Tracy)} by establishing that S = {@x: Girl(x) => Likes(Paul,x), Girl(Tracy)} and ~W = ~Likes(Paul,Tracy) are unsatisfiable. The two wffs in S imply Likes(Paul,Tracy) and this and ~W = ~Likes(Paul,Tracy) are unsatisfiable, implying a **contradiction**, and therefore that W = Likes(Paul,Tracy), in fact, logically follows from S.

An **inference procedure** derives new wffs from a given set of wffs. If all derived wffs — that is, all **inferences** — logically follow from the set, then the procedure, or the **derivation**, is **sound.** In Chapter 4, two sound inference procedures, binary resolution and binary factoring, are presented.

An inference procedure is **complete** if it will always find a proof to a theorem when given enough time.

## 2.5 A Set of Axioms to Prove Theorems in Group Theory

Researchers in automated theorem proving are often interested in proving theorems in group theory. A group G has an identity element e, an inverse element specified by the inverse function g(x) for every element x of the group, and a closure function f(x,y) that maps all pairs of group elements to a group element. The function f could be addition modulo some integer. Different sets of axioms are used to define a group. The one presented here uses two predicates: Sum(x,y,z) denotes "the sum of x and y is z," and Equal(x,y) denotes "x equals y." The axioms are:

The closure property of groups:
@x@y: Sum(x,y,f(x,y)) (G1)

The existence of an identity element (for example, 0 for addition):
@x: Sum(e,x,x) (G2)
@x: Sum(x,e,x) (G3)

Every element x has an inverse element g(x):
@x: Sum(g(x),x,e) (G4)
@x: Sum(x,g(x),e) (G5)

The associative laws:
@x@y@z@u@v@w: Sum(x,y,u) & Sum(y,z,v) & Sum(x,v,w) => Sum(u,z,w) (G6)
@x@y@z@u@v@w: Sum(x,y,u) & Sum(y,z,v) & Sum(u,z,w) => Sum(x,v,w) (G7)

The sum operation is well defined:
@x@y@z@w: Sum(x,y,z) & Sum(x,y,w) => Equal(z,w) (G8)

This last wff brings in the Equal **predicate**. In order for a theorem-proving program to reason with this important predicate, additional equality-specific axioms are necessary. The first three **equality axioms** pertain to equality's properties of reflexivity, symmetry, and transitivity. The remaining six axioms pertain to equality's substitution properties. These **equality substitution axioms** are necessary not only to reason in the domain of group theory but, as we see in the next section, in other domains as well.

Equality axioms:
@x: Equal(x,x) (G9)
@x@y: Equal(x,y) => Equal(y,x) (G10)
@x@y@z: Equal(x,y) & Equal(y,z) => Equal(x,z) (G11)

Equality substitution axioms for the function f:
@x@y@w: Equal(x,y) => Equal(f(x,w),f(y,w)) (G12)
@x@y@w: Equal(x,y) => Equal(f(w,x),f(w,y)) (G13)
@x@y: Equal(x,y) => Equal(g(x),g(y)) (G14)

Equality substitution axioms for the predicate Sum:
@x@y@z@w: Equal(x,y) & Sum(x,w,z) => Sum(y,w,z) (G15)
@x@y@z@w: Equal(x,y) & Sum(w,x,z) => Sum(w,y,z) (G16)
@x@y@z@w: Equal(x,y) & Sum(w,z,x) => Sum(w,z,y) (G17)

It might be pointed out that not all 17 axioms are necessary to characterize a group. The reader can use HERBY and THEO to show that, for example, G3 and G7 follow from the other axioms. In fact, G3, G7, G9, G10, G11, G12, G13, G14, G16, and G17 all follow from the subset of axioms G1, G2, G4, G5, G6, G8, and G15.

## 2.6 An Axiom System for Euclidean Geometry

Tarsky proposed an axiom system for Euclidean geometry in 1926 based on two predicates, between, B(x,y,z), and equidistance, D(x,y,z,u). More recently, Quaife (1989) made modifications, and his system is presented here. B(x,y,z) is interpreted as "y is between x and z." D(x,y,z,u) is interpreted as "the distance from x to y is equal to the distance from z to u." Variables x and y can be thought of as defining a line segment x-y, as can variables z and u. Five functions appear in the axioms: EXT (extension), IP (Inner Pasch), EUC1 (Euclid 1), EUC2 (Euclid 2) and CONT (continuity).

Reflexivity axiom for equidistance:
@x@y:D(x,y,y,x) (A1)

Transitivity axiom for equidistance:
@u@v@w@x@y@z: D(u,v,w,x) & D(u,v,y,z) => D(w,x,y,z) (A2)

Identity axiom for equidistance:
@u@v@w: D(u,v,w,w) => Equal(u,v) (A3)

Segment construction axiom:
@u@v@w@x: B(u,v,EXT(u,v,w,x)) (A4.1)
@u@v@w@x: D(v,EXT(u,v,w,x),w,x) (A4.2)

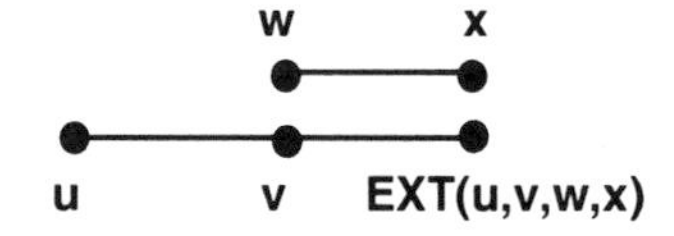


Outer five-segment axiom:
@u@v@w@x@u'@v'@w'@x': D(u,v,u',v') & D(v,w,v',w') & D(u,x,u',x') &
D(v,x,v',x') & B(u,v,w) & B(u',v',w') => Equal(u,v) I D(w,x,w',x') (A5)

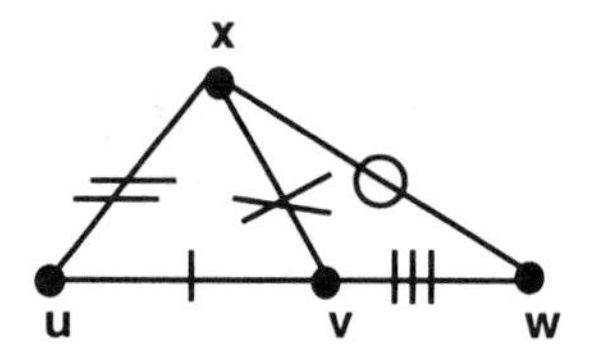

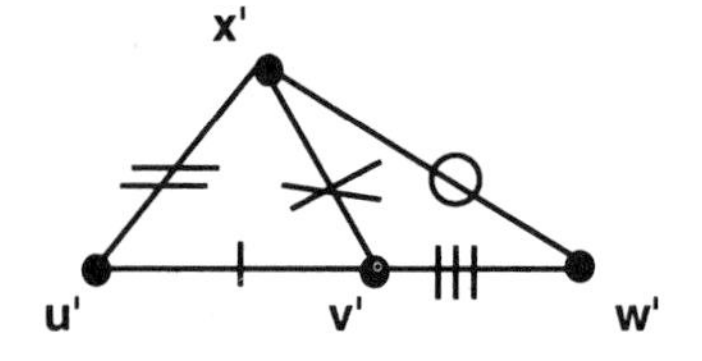

Identity axiom for betweenness:

@x@y: B(x,y,x) => Equal(x,y) (A6)

Inner Pasch axiom:

@u@v@w@x@y: B(u,v,w) & B(y,x,w) => B(v,IP(u,v,w,x,y),y) (A7.1)

@u@v@w@x@y: B(u,v,w) & B(y,x,w) => B(x,IP(u,v,w,x,y),u) (A7.2)

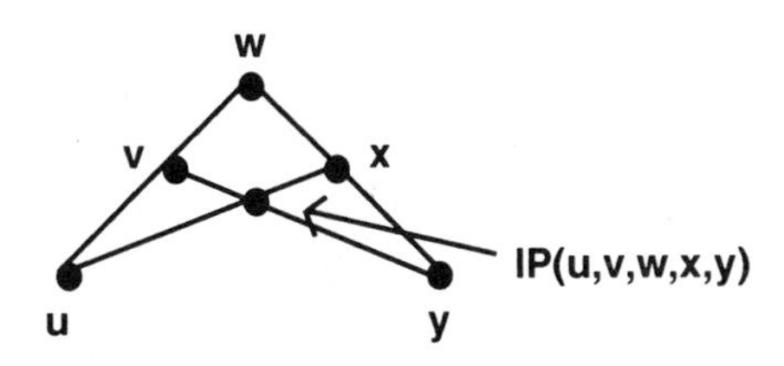


Lower dimension axiom (there exist three points not on the same straight line):

!x!y!z: ~B(x,y,z) & ~B(y,z,x) & ~B(z,x,y) (A8)

Upper dimension axiom:

@x@y@u@v@w: D(u,x,u,y) & D(v,x,v,y) & D(w,x,w,y) =>
Equal(x,y) I B(u,v,w) I B(v,w,u) I B(w,u,v) (A9)

Euclid's axioms:

@u@v@w@x@y: B(u,w,y) & B(v,w,x) =>
Equal(u,w) I B(u,v,EUC1(u,v,w,x,y)) (A10.1)

@u@v@w@x@y: B(u,w,y) & B(v,w,x) =>
Equal(u,w) I B(u,x,EUC2(u,v,w,x,y)) (A10.2)

@u@v@w@x@y: B(u,w,y) & B(v,w,x) =>
Equal(u,w) I B(EUC1(u,v,w,x,y),y,EUC2(u,v,w,x,y)) (A10.3)

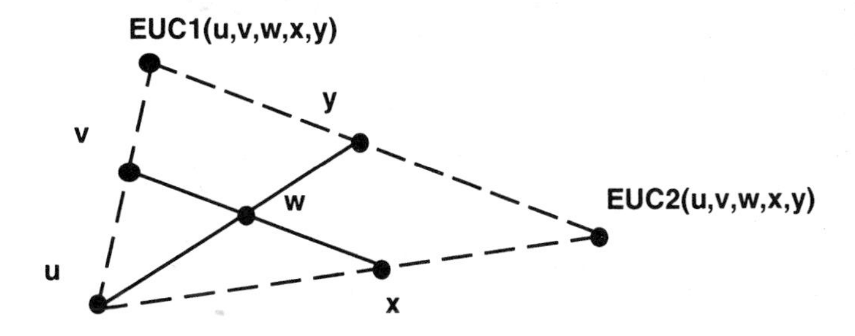

Weakened elementary continuity axiom (any segment that joins two points, one inside and one outside a given circle, intersects the circle):

@u@v@w@x@x'@v': D(u,v,u,v') & D(u,x,u,x') & B(u,v,x) & B(v,w,x) => B(v',CONT(u,v,v',w,x,x'),x') (A11.1)

@u@v@w@x@x'@v': D(u,v,u,v') & D(u,x,u,x') & B(u,v,x) & B(v,w,x) => D(u,w,u,CONT(u,v,v',w,x,x')) (A11.2)

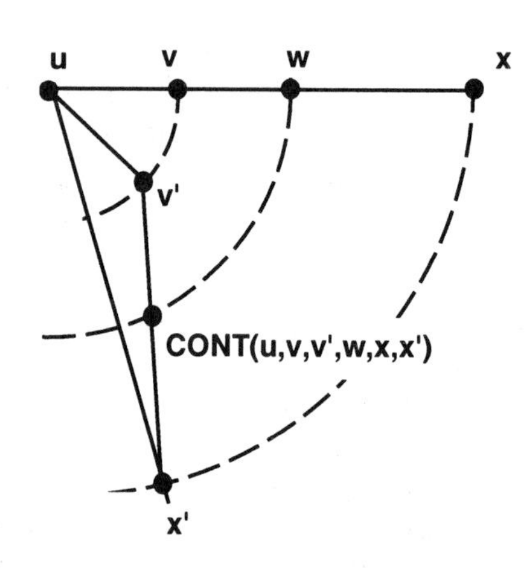


In addition to these 11 axioms, equality axioms and equality substitution axioms are also necessary:

Equality axioms:

@x: Equal(x,x) (E1)

@x@y: Equal(x,y) => Equal(y,x) (E2)

@x@y@z: Equal(x,y) & Equal(y,z) => Equal(x,z) (E3)

Equality substitution axioms for the predicate between:

@x@y@z@w: Equal(x,y) & B(x,z,w) => B(y,z,w) (BE1)

@x@y@z@w: Equal(x,y) & B(z,x,w) => B(z,y,w) (BE2)

@x@y@z@w: Equal(x,y) & B(z,w,x) => B(z,w,y) (BE3)

Equality substitution axioms for the predicate distance:

@x@y@z@u@v: Equal(x,y) & D(x,z,u,v) => D(y,z,u,v) (DE1)

@x@y@z@u@v: Equal(x,y) & D(z,x,u,v) => D(z,y,u,v) (DE2)

@x@y@z@u@v: Equal(x,y) & D(z,u,x,v) => D(z,u,y,v) (DE3)

@x@y@z@u@v: Equal(x,y) & D(z,u,v,x) => D(z,u,v,y) (DE4)

Equality substitution axioms for the function EXT:

@x@y@z@u@v: Equal(x,y) => Equal(EXT(x,z,u,v),EXT(y,z,u,v)) (XE1)
@x@y@z@u@v: Equal(x,y) => Equal(EXT(z,x,u,v),EXT(z,y,u,v)) (XE2)
@x@y@z@u@v: Equal(x,y) => Equal(EXT(z,u,x,v),EXT(z,u,y,v)) (XE3)
@x@y@z@u@v: Equal(x,y) => Equal(EXT(z,u,v,x),EXT(z,u,v,y)) (XE4)

Equality substitution axioms are also necessary for the functions IP, EUC1, EUC2, and CONT (see Exercise 2.5).

## Exercises for Chapter 2

2.1. Express the following statements as wffs.
(a) For all positive integers x, if x is a perfect square (i.e., 4, 9, 16, ...), then there exists a y such that y + 1 times y – 1 equals x – 1.
(b) For all positive integers x, if x is not a prime number, then there exists a y such that y is prime, y divides x, and y is less than x.
(c) Block C is above block B if block C is on block B or if block C is on some other block that is above block B.

2.2. Using as predicates hummingbirds, birds, richly_colored, large, and lives_on_honey, express the following syllogism as a set of wffs.
Axioms:
(a) All hummingbirds are richly colored.
(b) No large birds live on honey.
(c) Birds that do not live on honey are dull in color.
Conclusion:
(d) All hummingbirds are small.

2.3. Express the following syllogism as a set of wffs.
Axioms:
(a) None of the unnoticed things, met with at sea, are mermaids.
(b) Things entered in the log, and met with at sea, are worth remembering.
(c) I have never met anything worth remembering when on a voyage.
(d) Things met with at sea and noticed are recorded in the log.
Conclusion:
(e) I have never come across a mermaid at sea.

2.4. The axioms defining the natural numbers (i.e., 0, 1, 2, 3, . . . ) are:
(a) Every natural number has exactly one immediate successor.
(b) The natural number 0 is not the immediate successor of any natural number.
(c) Every natural number except 0 has exactly one immediate predecessor.
Convert these statements to wffs, using predecessor(x) and successor(x) as functions, and Equal(x,y) as the predicate.

2.5. Give equality substitution axioms for the functions for which these axioms were not given in Section 2.6.

2.6 Define the predicate collinear in terms of the predicate between, which was introduced in Section 2.6.

2.7. Using only the predicates between, distance, and Equal from Section 2.6, express the following theorem from Euclidean geometry as a wff: the diagonals of a nondegenerate rectangle bisect each other.

The following two exercises show how many problems can be posed as theorems. Exercises 2.8 and 2.9 are usually presented as state-space search problems, not as theorems. In Exercises 8.9 and 8.10, you are asked to find proofs to these theorems, equivalent to finding state-space solutions.

2.8. The missionaries and cannibals problem is a well-studied problem in artificial intelligence (AI):

> Imagine three missionaries and three cannibals stand on the left bank of a river. They all want to cross to the other side. They have a boat that can transport one or two of them at a time across the river. If, however, at any time the cannibals outnumber the missionaries on either side of the river, they will eat the missionaries. Prove: it is possible for all six to cross without a missionary being munched.

Express this problem as a theorem in first-order predicate calculus. (One predicate is sufficient. Let State_MCB(x,y,z) denote the state in which x missionaries are on the left bank, y cannibals are on the left bank, and the boat is on the z side, either left or right, of the river. The predicate State_MCB(3,3,L) is the initial state.)

2.9. The eight-puzzle is another well-studied problem in AI.

> Given an initial configuration of the eight-puzzle such as the one

following, the shortest sequence of moves of the blank square that transforms the initial state to the goal state is sought.

Initial State

| 1 | 3 | 4 |
|---|---|---|
| 7 | 2 | ■ |
| 6 | 8 | 5 |

Goal State

| 1 | 2 | 3 |
|---|---|---|
| 8 | ■ | 4 |
| 7 | 6 | 5 |

Express this problem as a theorem in first-order predicate calculus. Let the predicate State_Puzzle(x,y,z,u,v,w,X,Y,b) denote a configuration of the puzzle with the blank on the lower right-hand corner. The initial configuration corresponds to the predicate State_Puzzle(1,3,4,7,2,b,6,8,5), and the goal configuration corresponds to the predicate State_Puzzle(1,2,3,8,b,4,7,6,5). A wff corresponding to the possible "moves" when the blank is in the upper left-hand corner is:

```
@x@y@z@u@v@w@X@Y: State_Puzzle(b,x,y,z,u,v,w,X,Y) =>
            State_Puzzle(z,x,y,b,u,v,w,X,Y) & State_Puzzle(x,b,y,z,u,v,w,X,Y)
```

For each of the other eight squares, write a wff such as the preceding one for the upper left-hand corner.

THEO
31
19
3
59
2
17
5
13

# 3 COMPILE: Transforming Well-Formed Formulas to Clauses

Chapter 2 explained how to write a theorem as a set of wffs. However, neither HERBY nor THEO attempt to find a proof of a theorem expressed in this format. Instead, both require the theorem to be expressed as a set of **clauses** consisting of the axioms, hypotheses, and negated conclusion. Then, using the clauses as input, an attempt is made to find a proof. The program COMPILE transforms wffs to clauses by carrying out the seven steps described in Section 3.1. Section 3.2 explains how to use COMPILE, with the actual details explained in Section 3.2.1. Examples are given in Section 3.2.2. It should be pointed out that the user can express a theorem as a set of clauses directly, circumventing the need to use COMPILE.

## 3.1 The Transformation Procedure of COMPILE

The following wff serves as a working example (see EXCOMP1.WFF):

@x: {@y: P(a,y) & S(y,x) } => ~{@y: Q(x,y) & S(y,x) => R(x,y)}

Step 1. Eliminate the operators =>, and <=> by replacing them as follows:

Replace A => B with ~A | B
Replace A <=> B with {~A | B} & {A | ~B}

For example, replace @x: P(x) => Q(x) with @x: ~P(x) | Q(x). Applying Step 1 to the working example gives:

@x: ~{@y: P(a,y) & S(y,x) } | ~{@y: ~{Q(x,y) & S(y,x)} | R(x,y)}

Step 2. Distribute negation symbols over other logical symbols until each negation symbol applies directly to a single predicate. In particular:

| | | | |
|---|---|---|---|
| Replace | ~~A | with | A |
| Replace | ~{A I B} | with | ~A &~B |
| Replace | ~{A & B} | with | ~A I ~B |
| Replace | ~{@x: A} | with | {!x: ~A} |
| Replace | ~{!x: A} | with | {@x: ~A} |

Applying Step 2 to the example gives:

@x: {!y: ~P(a,y) I ~S(y,x)} I {!y: Q(x,y) & S(y,x) & ~R(x,y)}

Step 3. Rename variables so that no two quantifiers quantify the same variable. For example, replace ~{@x: P(x)} I {!x: Q(x)} with ~{@x: P(x)} I {!y: Q(y)}. Applying Step 3 to the working example gives:

@x: {!y: ~P(a,y) I ~S(y,x)} I {!z: Q(x,z) & S(z,x) & ~R(x,z)}

Step 4. For each existentially quantified variable x, eliminate the !x (and the colon if appropriate) and then replace the variable wherever it appears in the expression with either a Skolem constant or a Skolem function as explained in the next two paragraphs and exemplified in Figure 3.1. There are two cases to consider.

(a) If the !x is not within the scope of any universal quantifier, replace each occurrence of x with a new constant — a Skolem constant — that does not appear elsewhere in the wff and that corresponds to an instance of the x that

Example 1. Replace !x@y!z@w: P(x,y,z,w)
with @y@w: P(SK1,y,SK2(y),w)

Example 2. Replace {@x@y!z: P(x,f(y,z),z} I !w: R(w)
with {@x@y: P(x,f(y,SK1(x,y)),SK1(x,y))} I R(SK2)

Example 3. Replace @x!y: P(x,y) I {@z!w: Q(x,y,z,w)}
with @x: P(x,SK1(x)) I {@z: Q(x,SK1(x),z,SK2(x,z))}

Example 4. Replace @y: {!w: P(y,w) I Q(w)} I {@u!z: R(y,z,u)}
with @y: {P(y,SK1(y)) I Q(SK1(y)} I {@u: R(y,SK2(y,u),u)}

Figure 3.1. Four examples illustrating Step 4.

exists. For example, replace !x: P(x) with P(SK), where SK is a Skolem constant. Because SK corresponds to an x that exists in !x: P(x), it follows that !x: P(x) implies P(SK) and vice versa. Two other examples are: (1) replace !x@y: P(x,y,x) with @y: P(SK,y,SK), and (2) replace !x!y@z: P(x,y,z) with @z: P(SK1,SK2,z).

(b) If the !x is within the scope of one or more universal quantifiers, u, v, w, ..., replace each occurrence of x with a Skolem function (1) that does not appear elsewhere in the wff and (2) whose arguments are the universally quantified variables u, v, w, ... and (3) which evaluates to the x that exists for the corresponding values of u,v,w, .... For example, replace @x@y!z: P(x,y,z) with @x@y: P(x,y,SK(x,y)), where SK(x,y) is a Skolem function whose arguments are the universally quantified variables x and y. It follows that @x@y!z: P(x,y,z) implies @x@y: P(x,y,SK(x,y)) if for all pairs x and y the Skolem function SK(x,y) corresponds to an instance of z that exists and vice versa.

Applying Step 4 to the working example gives:

@x: {~P(a,SK1(x)) | ~S(SK1(x),x)} | {Q(x,SK2(x)) & S(SK2(x),x) & ~R(x,SK2(x))}

Step 5. At this point, all variables in the expression are universally quantified. Dropping the @x@y@z... does not change the meaning of the expression if this is understood.

Applying Step 5 to the working example gives:

{~P(a( ),SK1(x)) | ~S(SK1(x),x)} | {Q(x,SK2(x)) & S(SK2(x),x) & ~R(x,SK2(x))}

Note that COMPILE adds parentheses to the end of the names of constants here because there are now no universal quantifiers to distinguish between constants and variables now. A constant is essentially a function with no arguments.

Step 6. Convert the results of Step 5 to conjunctive normal form; that is, to a Boolean conjunction of disjunctions of literals. Each disjunction is called a **clause**. This conversion is done by repeatedly applying the distributive rules: replace A | {B & C} with {A | B} & {A | C}, and replace {A & B} | C with {A | C} & {B | C}.

Applying Step 6 to the working example gives:

{~P(a( ),SK1(x)) I ~S(SK1(x),x) I Q(x,SK2(x))}
& {~P(a( ),SK1(x)) I ~S(SK1(x),x) I S(SK2(x),x)}
& {~P(a( ),SK1(x)) I ~S(SK1(x),x) I ~R(x,SK2(x))}

Step 7. Rewrite the results of Step 6 as a set of clauses, one to a line, where it is now understood that the logical & is between the clauses. For example, replace {P I Q} & {R I S} & {R I S I T} with the three clauses:

P I Q
R I S
R I S I T

where it is understood that the logical & is between the clauses.

Applying Step 7 to the working example gives:

~P(a( ),SK1(x)) I ~S(SK1(x),x) I Q(x,SK2(x))
~P(a( ),SK1(x)) I ~S(SK1(x),x) I S(SK2(x),x)
~P(a( ),SK1(x)) I ~S(SK1(x),x) I ~R(x,SK2(x))

The variable x that appears in each of these three expressions could be renamed so that the names of variables in each clause were distinct. This step is not performed by COMPILE.

A set of clauses created by these seven steps is satisfiable if and only if the original wff is satisfiable. In other words, these seven steps do not affect satisfiability.

## 3.2 Using **COMPILE**

COMPILE transforms a set of wffs, which must be in a text file, to wffs. The wffs must be constructed according to the rules presented in Section 2.1. In addition, each wff must be terminated with a period (.). The text file of wffs can contain comments. Any text following a semicolon in a line is considered a comment. A wff can be longer than one line; carriage returns are ignored. The key word "conclusion" tells COMPILE that any wffs that follow are part of the conclusion. COMPILE negates these wffs before transforming them to clauses. Figure 3.2 illustrates how COMPILE works.

To transform a file of wffs, say AGATHA.WFF, type the command line:

COMPILE

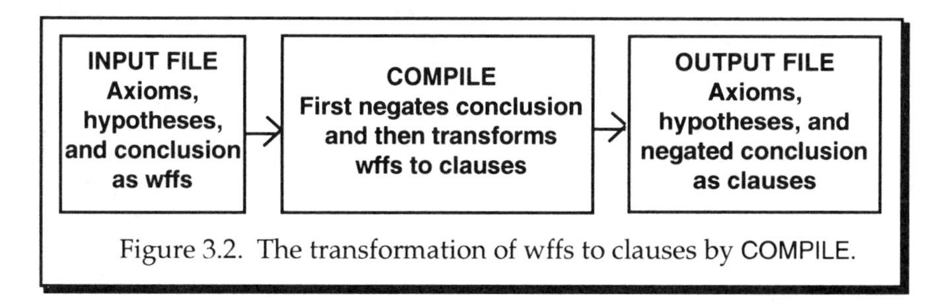


Figure 3.2. The transformation of wffs to clauses by COMPILE.

The computer responds with:

Enter the name of the theorem:

You should then enter the name of the file that contains the wffs:

WFFS/AGATHA.WFF

The transformed clauses are displayed on the screen and simultaneously saved in the active directory in the file AGATHA.T. Those clauses following the key word "negated_conclusion" constitute the negated conclusion.

## 3.3 Examples of the Transformation of Wffs to Clauses

Four examples of the transformation of wffs to clauses are presented in Figure 3.3. The examples are in the directory WFFS.

## Exercises for Chapter 3

3.1. Convert the following wffs to clauses. The last four are too difficult to do by hand, so try COMPILE. The last two are too difficult even for COMPILE!

(a) @x@y: E(x,y) => E(y,x)

(b) @x: A(x) & B(x) & C(x) => D(x)

(c) @x: A(x) => B(x) & C(x) & D(x)

Example 1. The wff (see EXCOMP2.WFF):

```
@x@y: woman(x) & man(y) & poet(y) => likes(x,y)
```

transforms to:

```
~woman(x) | ~man(y) | ~poet(y) | likes(x,y)
```

Example 2. The wff (see EXCOMP3.WFF):

```
@x: {@y: P(x,y)} => ~{@y: Q(x,y) => R(x,y)}
```

transforms to:

```
~P(x,SK1(x)) | Q(x,SK2(x))
~P(x,SK1(x)) | ~R(x,SK2(x))
```

Example 3. The wff (see EXCOMP4.WFF):

```
@x: P(x) => {@y: P(y) => P(f(x,y))} & ~{@y: Q(x,a,y) => P(y)}
```

transforms to:

```
~P(x) | ~P(y) | P(f(x,y))
~P(x) | Q(x,a( ),SK1(x))
~P(x) | ~P(SK1(x))
```

Example 4. The wff (see EXCOMP5.WFF):

```
~{@x: P(x) => {@y: P(y) => P(f(x,y))} & ~{@y: Q(x,a,y) => P(y)}}
```

transforms to:

```
P(SK1( ))
~Q(SK1( ),a(),x) | P(x) | P(SK2( ))
~Q(SK1( ),a(),x) | P(x) | ~P(f(SK1( ),SK2( )))
```

Figure 3.3. Examples of the transformation of wffs to clauses.

(d) @x: A(x) => B(x) | C(x) | D(x)

(e) @x: A(x) | B(x) | C(x) => D(x)

(f) @x: A(x) => {B(x) => C(x)} (see EXER3F.WFF)

(g) @x: P(x) => {@y: P(y) =>
P(f(x,y))} | ~{@y: {Q(x,a,y) => P(y)}} (see EXER3G.WFF)
(Note: a is a constant.)

(h) @x@y: set(x) & set(y) => {Equal(x,y) <=>
{@u: in(u,x) <=> in(u,y)}} (see EXER3H.WFF)

(i) @x@y: P(x,y) | S(y,x) =>
~{@z: Q(x,z) & S(z,x) => R(x,z)} (see EXER3I.WFF)

(j) !x@y@z!w@u!v: P(x,y,z) => Q(w,u) & R(v,x,v) (see EXER3J.WFF)

(k) ~{y&~z | ~x&y => y&~z | ~x&y&z} (see EXER3K.WFF)

(l) ~{y&~z | ~x&y | ~x&z&~
w <=> y&~z | ~x&y&z | ~x&~y&z&~w} (see EXER3L.WFF)

(m) ~{{y&~z} | {~x&y} | {~x&z&~w} => {y&~z} |
{~x&y&z} | {~x&~y&z&~w}} (see EXER3M.WFF)

(n) ~{{{y&~z} | {~x&y} | {~x&z&~w} <=> {y&~z} |
{~x&y&z} | {~x&~y&z&~w}}} (see EXER3N.WFF)

(o) {{!x@y: P(x) <=> P(y) } <=> {!x: Q(x) <=> {@y: Q(y)}}}
<=> {{!x@y: Q(x) <=> Q(y)} <=> {!x: P(x) <=> {@y: P(y)}}}
(see EXER3O.WFF)

3.2. Convert the wffs corresponding to the axioms for natural numbers given in Exercise 2.4 to clauses.

3.3. Convert the wffs corresponding to the group theory axioms G1 through G17 given in Section 2.5 to clauses.

3.4. Convert the wffs corresponding to the Euclidean geometry axioms A1 through A11 given in Section 2.6 to clauses.

3.5. Prove that @x: P(x) => !y: P(y) by negating the wff and then showing that this implies a contradiction.

3.6. Negate the following wff and then convert the resulting wff to clause form: @x@y@z@w: B(x,y,z) & B(y,w,z) => B(x,w,z).

3.7. The theorem STARK017.THM says, essentially, that there exists an infinite number of prime numbers. The theorem is presented as a set of clauses with the conclusion negated. Express as English statements the axioms, hypothesis, and conclusion. Note that F(x) = x! + 1, and that H(x) is a Skolem function.

3.8. [Project] Examine the clauses produced by COMPILE for the more difficult exercises; note that they can be simplified considerably. The shortcoming with COMPILE is the conversion of an expression to conjunctive normal form. It does not check to see if a tautology has been produced and can be eliminated, or if two identical literals exist in a clause, in which case one can be eliminated. Further, it does not eliminate subsumed clauses (see Section 4.9). Modify the source code to correct these deficiencies. Can your modified version transform all of the wffs presented in this chapter to clause form? You might refer to Socher's (1991) paper listed in the references.

3.9. [Project] Improve the interactive capabilities of COMPILE by giving the user options to observe the compilation steps as they are carried out.

# 4 Inference Procedures

Both HERBY and THEO use only two inference procedures to derive new clauses from a given set of clauses; the procedures are called binary resolution and binary factoring. To understand how these procedures derive new clauses, a number of terms must be introduced — in particular, substitution, instance, unification, subsumption and most general unifier. An informal introduction to this material, however, is given first.

To this point in the text, the names of constants have been followed by parentheses. In the remainder of this book, these parentheses will be omitted; variable names usually will be chosen from letters near the end of the alphabet and constants from the beginning. In addition, names such as "bugs" and "vegetables" will be those of constants.

## 4.1 An Informal Introduction to Binary Resolution and Binary Factoring

First, we introduce binary resolution: If A => B and B => C, then it follows that A => C. The process of binary resolution is a generalization of this rule of logic to the case where the literals are not simply predicates with no arguments and where A, B, and C are not restricted to being single literals. Consider the two clauses C1 and C2:

```
C1: ~bunny(bugs) | likes(bugs,vegetables)
C2: ~likes(x,vegetables) | likes(x,carrots)
```

They correspond to the following pair of inferences:

```
C1': bunny(bugs) => likes(bugs,vegetables)
C2': likes(x,vegetables) => likes(x,carrots)
```

If bugs is substituted for x in C2, the second literal of C1 becomes identical to the first literal of C2 — it will be learned later that the two literals have a common instance — and it then follows from C1 and C2 that:

C3: bunny(bugs) => likes(bugs,carrots)

or as a clause:

C3: ~bunny(bugs) | likes(bugs,carrots)

C3 is a binary resolvent of C1 and C2. Although C1 and C2 have only one binary resolvent, more generally two clauses can have many binary resolvents. For example, the two clauses:

C4: bunny(bugs) | bunny(roger)
C5: ~bunny(x) | likes(x,carrots)

have two resolvents:

C6: bunny(bugs) | likes(roger,carrots)
C7: bunny(roger) | likes(bugs,carrots)

These two resolvents are somewhat harder to comprehend than the original ones, and this is often the case. Randomly resolving clauses together does not usually produce new clauses that are more understandable than the original ones. Resolving two clauses together can yield a contradiction: consider the following two **unit clauses**; that is, clauses with a single literal:

C8: ~bunny(x)
C9: bunny(bugs)

If the substitution bugs for x is made in C8, the two unit clauses become contradictory; C8 becomes ~bunny(bugs) and C9 remains bunny(bugs). These two contradictory clauses resolve to the NULL **clause**, denoted by **Ø**. Unit clauses are particularly important to THEO: the last step in every proof is the resolution of two contradictory unit clauses.

Binary factoring is a generalization of the rule of logic, A | A implies A, to the case where A is not simply a predicate with no arguments. As a simple example, consider the clause less_than(2,x) | less_than(y,10). A binary factor of this clause is less_than(2,10), an instance common to both literals. More complex examples of factoring are possible and are examined later in this chapter.

A formal description of the procedure of binary resolution and binary factoring requires an understanding of the processes of substitution and unification.

## 4.2 The Processes of Substitution and Unification

Consider the literal likes(x,y). likes(jane,y) is called an **instance** of likes(x,y). More specifically, likes(x,y) is a **substitution instance** of likes(x,y) resulting from the substitution of the constant jane for the variable x. This substitution, denoted by s1, can be written as: s1 = {jane/x}. The substitution instance likes(joe,jane) is called a **ground instance** of the literal likes(x,y), resulting from the substitution of the constants joe and jane for x and y, respectively, because the instance has no variables. A **ground clause** is a clause in which each literal is a ground instance.

More formally, a **substitution** {t1/v1, t2/v2, ..., tn/vn} is a list of terms t1, t2, ..., tn and a list of variables v1, v2, ..., vn, which are **bound** or assigned the values of the corresponding terms with the restriction that no bound variable can appear in any term. The terms are called the **bindings** of the variables. When the substitution is applied to a literal, producing a substitution instance of that literal, each occurrence of v1 is replaced by t1, each occurrence of v2 is replaced by t2, and so on. For the **NULL substitution**, denoted {ø/ø}, no variables are replaced.

Consider the literal P(x,y,f(x,z),w). P(a,y,f(a,g(b)),g(c)) is a substitution instance of P(x,y,f(x,z),w) resulting from the substitution of a for x, g(b) for z, and g(c) for w. This substitution, denoted by s2, can be written as:

s2 = {a/x,g(b)/z,g(c)/w}

Note that, in the preceeding examples, terms are substituted for variables. The application of substitutions s1 and s2 can be described by:

likes(jane,y) = [likes(x,y)]s1
P(a,y,f(a,g(b)),g(c)) = [P(x,y,f(x,z),w)]s2

More generally, a substitution can be defined as an operation on a clause—not simply on a literal—and in this case, the substitution is applied to each literal in the clause. For example, when the substitution{28/x,7/y} is applied to the clause ~divisible_by(x,y) I factor_of(y,x), the substitution instance is ~divisible_by(28,7) I factor_of(7,28).

Now, consider two instances P1 and P2 of the same predicate. Let some substitution $\pi$ be applied to both instances. Denote the resulting instances of P1 and P2 by P1' and P2', respectively. According to the preceding notation, P1' = [P1]$\pi$ and P2' = [P2]$\pi$. Now, if P1' = P2' = P, P1 and P2 are said to **unify** to a **common instance** P and $\pi$ is the **unifier**. Figure 4.1 presents five examples of unification.

Note that P1 = P(x,x) and P2 = P(y,f(y)) have no common instance. They

Example 1. likes(bugs,x) and likes(y,carrots) are unified by the substitution {bugs/y, carrots/x} to likes(bugs,carrots).

Example 2. P(x,x) and P(z,a) are unified by the substitution {a/z,a/x} to P(a,a).

Example 3. likes(x,daughter_of(molly,sam)) and likes(jack,y) are unified by the substitution {jack/x,daughter_of(molly,sam)/y} to likes(jack,daughter_of(molly,sam)).

Example 4. Q(x,f(x,y),z) and Q(a,u,v) are unified by the substitution {a/x,f(a,y)/u,z/v} to Q(a,f(a,y),z).

Example 5. R(x,f(x,y),g(f(a,y))) and R(b,f(u,c),z) are unified by the substitution {b/x,b/u,c/y,g(f(a,c))/z} to R(b,f(b,c),g(f(a,c))).

Figure 4.1. Unification examples.

do not unify because there is no unifying substitution. Substituting x for y yields P1' = P(y,y) and P2' = P(y,f(y)) and P1 ≠ P2. Substituting f(y) for x yields P1' = P(f(y),f(y)) and P2' = P(y,f(y)); again P1' ≠ P2'. No other substitution is possible.

Let π and ß be two substitutions. ß is said to be distinct from π if and only if no variable bound in π appears in ß. (A variable that appears in a binding in π can appear in ß.) The **composition** of a distinct π with ß and denoted πß is the substitution that results by applying ß to the terms of π and adding the bindings from ß. Only compositions of distinct ß with π are made in the material that follows. For example, if π = {f(x,y)/z} and ß = {a/x,b/y,u/w}, then πß = {f(a,b)/z,a/x,b/y,u/w}. Note that, in general, ßπ ≠ πß.

## 4.3 Subsumption

Suppose P1 and P2 are two instances of predicate P. P1 is said to **subsume** P2 if there is a substitution π such that [P1]π = P2. For example, suppose P1 = likes(x,y) and P2 = likes(fred,sue). Then, likes(fred,sue) is subsumed by likes(x,y) because if π = {fred/x,sue/y}, then [P1]π = P2. Figure 4.2 shows three examples of subsumption. The concept of subsumption is extended to clauses in Section 4.8.

Example 1. P(x,y,z) subsumes P(u,a,a) with substitution {u/x,a/y,a/z}.

Example 2. P(x,y,b) subsumes P(z,g(w),b) with substitution {z/x,g(w)/y}.

Example 3. P(x,g(u),f(v)) subsumes P(w,g(a),f(h(y,z))) with substitution {w/x,a/u,h(y,z)/v}.

Figure 4.2. Examples of subsumption.

## 4.4 The Most General Unifier

There are often a number of substitutions that unify two instances of the same predicate. For example, P(x,y,a) and P(z,z,w) unify under the substitution ß = {a/x,a/y,a/z,a/w} to P(a,a,a) and under the substitution ∂ = {z/x,z/y,a/w} to P(z,z,a). ∂ produces a substitution instance that subsumes the substitution instance produced by ß. That is, ∂ produces a more common substitution instance than does ß. The **most general unifier (mgu)**, μ, of two predicate instances P1 and P2 is defined as the unifier that produces a substitution instance P3 such that P3 subsumes every other substitution instance of P1 and P2, that is, that produces the **most common instance**.

An algorithm for calculating the mgu of two instances of predicate P is illustrated in Figure 4.3, where P1 = P(x,f(x,y),g(f(a,z))) and P2 = P(u,f(g(a),u),w) serve as a working example. Pointers are initially set to the leftmost symbols inside the first parenthesis. If the symbols match, the pointers are moved right to the first unmatched pair of symbols, if such a pair exists. If the symbols are the names of two different functions, no substitution is possible and no mgu exists. Otherwise, the term t pointed to in one literal is substituted for the variable v pointed to in the other — one of the symbols pointed to must be a variable. If, however, t contains v, then the substitution leads to an infinite loop, and no mgu exists. The pointers are next moved right to the next unmatched pair of symbols. If the pointers reach the right end of the predicate, an mgu exists. In Figure 4.4, the mgu is given for three examples.

## 4.5 Determining All Binary Resolvents of Two Clauses

The disjunction of clauses C1 and C2 is the clause C3 = C1 | C2, where the "|" symbol denotes disjunction. For example, if C1 = P(x,y) | Q(y,z) and C2 = ~P(a,w) | T(u), then C3 = C1 | C2 = P(x,y) | Q(y,z) | ~P(a,w) | T(u).

Consider calculating the mgu of the two predicates, P1 and P2:

```
P1 = P(x,f(x,y),g(f(a,z)))          P2 = P(u,f(g(a),u),w)
       ^                                   ^
```

First substitution {u/x}: substitute u for x (or x for u). This gives, with the pointers moved to the next unmatched symbols:

```
P1 = P(x,f(x,y),g(f(a,z)))          P2 = P(x,f(g(a),x),w)
         ^                                   ^
```

Second substitution {g(a)/x}. The composition of the first two substitutions is {g(a)/u,g(a)/x}. This transforms P1 and P2 to:

```
P1 = P(g(a),f(g(a),y),g(f(a,z)))          P2 = P(g(a),f(g(a),g(a)),w)
                   ^                                         ^
```

Third substitution {g(a)/y}. The composition of {g(a)/u, g(a)/x} and {g(a)/y} is {g(a)/u, g(a)/x, g(a)/y}. This transforms P1 and P2 to:

```
P1 = P(g(a),f(g(a),g(a)),g(f(a,z)))          P2 = P(g(a),f(g(a),g(a)),w)
                         ^                                          ^
```

Fourth substitution {g(f(a,z))/w}. The composition of {g(a)/u,g(a)/x,g(a)/y} and {g(f(a,z)/w} is {g(a)/u,g(a)/x,g(a)/y,g(f(a,z))/w}, the mgu of the two predicates. This substitution made to the original literals yields:

P1 = P2 = P(g(a),f(g(a),g(a)),g(f(a,z)))

Figure 4.3. Calculating the mgu of two literals.

Example 1. P(x,f(x),f(a)) and P(f(y),z,u) have the mgu {f(y)/x,f(f(y))/z,f(a)/u}. The predicates unify to P(f(y),f(f(y)),f(a)).

Example 2. P(u,v,u,v,f(a)) and P(f(x),f(y),z,x,y) have the mgu {f(f(f(a)))/u,f(f(a))/v,f(f(f(a)))/z,f(f(a))/x,f(a)/y}. The predicates unify to P(f(f(f(a))),f(f(a)),f(f(f(a))),f(f(a)),f(a)).

Example 3. P(f(x,g(x)),z) and P(f(g(y),v),h(a)) have the mgu {g(y)/x,g(g(y))/v,h(a)/z}. The literals predicates to P(f(g(y),g(g(y))),h(a)).

Figure 4.4. The mgu of three pairs of literals.

Suppose C is a clause and L is a literal in C; then C' = C – L denotes the clause that results from deleting literal L from clause C. For example, if C = P(x,y) | Q(x,f(x)) | ~R(g(x)) and L is Q(x,f(x)), then C' = C – L = P(x,y) | ~R(g(x)).

Consider two clauses C1 = L11 | L12 | ... | L1n and C2 = L21 | L22 | ... | L2m. Suppose some literal L1i appears in C1 and some literal L2j appears in C2 such that L1i and L2j are instances, possibly the same or different, of the same predicate P, complemented in one clause and uncomplemented in the other, and suppose the two instances unify with some mgu $\mu$. Then, C1 and C2 **resolve together** to form

C3: [{C1 – L1i} | {C2 – L2j}]$\mu$

which is called a **binary resolvent** or, more simply, a **resolvent** of C1 and C2. It is formed by **resolving on** L1i and L2j. Clauses C1 and C2 are called the **input clauses** to the resolvent. Literals L1i and L2i are sometimes said to be **resolved away**.

For example, suppose C4 and C5 are:

C4: P(x,y) | Q(x,z)
C5: ~Q(u,b) | ~P(u,a)

Because P(x,y) and P(u,a) unify with mgu {x/u,a/y}, one binary resolvent can be formed by resolving on P(x,y) in C4 and on ~P(u,a) in C5 forming:

C6: [P(x,y) | Q(x,z) – P(x,y) | ~Q(u,b) | ~P(u,a) – ~P(u,a)]{x/u,a/y}

which becomes, after the complementary literals are deleted, the substitutions made, and the variables normalized (defined in the next paragraph):

C7: Q(x,y) | ~Q(x,b)

Because all variables in a clause are universally quantified, their names can be changed without changing the meaning of the clause. When forming a binary resolvent of two clauses, THEO and HERBY begin by changing the names of the variables appearing in the two clauses to ensure that they are different — that is, the variables are **standardized apart**. (If the variables are already different, this step is unnecessary.) They next form the resolvent and then rename variables in the resolvent in the following **normal order**: x, y, z, u, v, w, t, s, r, q, p, o, n, m, l, k, K, L, M, N, O, P, Q, R, S, T, W, V, U, Z, Y, X. (A clause can have at most 32 variables.) In other words, the first variable, from the left-hand side, is called x, the next is called y, and so on.

C7 is formed by resolving on literal a of C4 and literal b of C5, where literals are indexed alphabetically from the beginning of the clause. In the notation used by THEO and HERBY and in the notation used henceforth in this text:

C7: (4a,5b) Q(x,y) I ~Q(x,b)

A second binary resolvent of C4 and C5 is

C8: (4b,5a) P(w,y) I ~P(w,a)

To determine all binary resolvents of two clauses, it is necessary to try to resolve together all literal pairs having one literal uncomplemented in one clause and the other literal complemented in the other clause. For each such pair that resolve together, there is a binary resolvent. Consider clauses C9 and C10:

C9: P(x,f(a,y)) I Q(g(b),y) I ~R(x,z,a) I ~R(x,f(x,x),a)
C10: ~P(g(b),f(u,v)) I ~Q(v,g(u)) I R(u,u,a)

These clauses have three binary resolvents. The first is formed by resolving on the first literal of each clause:

C11: [C9 – P(x,f(a,y)) I C10 – ~P(g(b),f(u,v))] {g(b)/x,a/u,v/y}

or:

C11: (9a,10a) = Q(g(b),x) I ~Q(x,g(a)) I ~R(g(b),y,a)
I ~R(g(b),f(g(b),g(b)),a) I R(a,a,a)

The second is formed by resolving on the second literal of each clause:

C12: [{C9 – Q(g(b),y)} I {C10 – ~Q(v,g(u))}]{g(b)/v,g(u)/y}

or:

C13: (9b,10b) P(x,f(a,g(y))) I ~P(g(b),f(y,g(b)))
I ~R(x,z,a) I ~R(x,f(x,x),a) I R(y,y,a)

The third is formed by resolving on the third literal of each clause:

C14: [{C9 – ~R(x,z,a)} I {C10 – R(u,u,a)}]{x/u,x/z}

or:

C15: (9c,10c) P(x,f(a,y)) I ~P(g(b),f(x,z))
I Q(g(b),y) I ~Q(z,g(x)) I ~R(x,f(x,x),a)

## 4.6 Merge Clauses

Suppose two or more identical instances of a literal appear in the binary resolvent of two clauses. All but one of these identical instances can be deleted without changing the meaning of the binary resolvent. The resulting binary resolvent in which these literals have been "merged together" is called a **merge clause**. For example, suppose C16 and C17 are:

C16: P(a,x) I P(y,b) I Q(x) I R(a,x,y)
C17: ~P(u,v) I Q(v) I ~R(v,b,a)

Note that variables in C17 have different names from those in C16. Otherwise, they should initially be standardized apart. Then, the binary resolvent:

C18: (16a,17a) P(x,b) I Q(y) I R(a,y,x) I Q(y) I ~R(y,b,a)

has two identical instances of Q(y). These two identical instances are merged together, yielding a merge clause:

C19: (16a,17a) P(x,b) I Q(y) I R(a,y,x) I ~R(y,b,a)

Clauses C16 and C17 have a second binary resolvent that is a merge clause:

C20: (16d,17c) P(a,b) I P(a,b) I Q(b) I ~P(x,a) I Q(a)

or:

C21: (16d,17c) P(a,b) I Q(b) I ~P(x,a) I Q(a)

Note that if C22 = P(x) I Q(y) and C23 = P(w) I ~Q(z), the two literals of the resolvent C24: (22b,23b) P(x) I P(w) must both be retained.

There are thus two types of binary resolvents: those that are merge clauses and those that are not. Merge clauses will be seen in Chapters 6 and 9 to have a special significance in the search for a proof.

## 4.7 Determining All Binary Factors of a Clause

Suppose clause C contains two instances of some literal either with both instances complemented or both instances uncomplemented. If the two instances unify with mgu $\mu$, then the clause C' = $[C]\mu$ contains two identical literals, and one can be deleted without changing the meaning of C'. The resulting clause is called a **binary factor** of C.

For example, consider the clause C25:

C25: likes(x,carrots) I likes(bugs,y)

likes(x,carrots) and likes(bugs,y) have an mgu {bugs/x,carrots/y}. If this substitution is made, clause C25 becomes:

C26: likes(bugs,carrots) I likes(bugs,carrots)

Deleting the second likes(bugs,carrots) gives the factor of clause C25:

C27: likes(bugs,carrots)

Binary factoring operates on two literals. In general, factoring is defined as an operation on any number of literals — what might be called **multiple factoring.** Of course, any multiple factor can be produced by performing a sequence of binary factorings. As a second example, consider the clause C28:

C28: P(x,y) I P(a,z) I P(x,b) I Q(a,y,x) I Q(z,b,z)

It has four binary factors:

C29: (28ab) P(a,x) I P(a,b) I Q(a,x,a) I Q(x,b,x)
C30: (28ac) P(x,b) I P(a,y) I Q(a,b,x) I Q(y,b,y)
C31: (28bc) P(a,x) I P(a,b) I Q(a,x,a) I Q(b,b,b)
C32: (28de) P(a,b) I P(a,a) I Q(a,b,a)

In producing C32, the literal P(a,b) occurred twice; one was deleted, producing a merge clause.

Finding all binary factors of a clause requires scanning the clause for all pairs of literals that either have both literals complemented or both uncomplemented. An mgu is sought for each such pair, and if one is found, a binary factor can be produced.

## 4.8 A Special Case of Binary Resolution: Modus Ponens

A special case of binary resolution called **modus ponens** is the case in which two clauses A and ~A I B imply the clause B. Alternatively, the two wffs A => B and A imply the wff B.

## 4.9 Clauses and Subsumption

The concept of subsumption, discussed in Section 4.3 in the context of two literals, can be extended to clauses. A clause C1 is said to **subsume** a clause C2 if there is a substitution ∂ such that the literals of [C1]∂ are a subset of the literals of C2.

To determine whether one clause subsumes another is a computationally intensive procedure and is not done in either THEO or HERBY. THEO tests instead for **simple subsumption**: Clause C1 simply subsumes or **s-subsumes** C2 if C1 subsumes C2 and the ∂ given in the previous paragraph is the NULL substitution and the number of literals in C1 is less than or equal to the number of literals in C2. It is computationally much faster to determine whether one clause is s-subsumed by another than whether it is more generally subsumed. Examples illustrating subsumption and s-subsumption are given in Figure 4.6.

Note that if some clause C1 subsumes some other clause C2, then C1 implies C2. For example, because Q(x) subsumes P(x) | Q(y), it follows that Q(x) implies P(x) | Q(y); that is, as a wff, {@x: Q(x)} => {@x@y: P(x) | Q(y)}. When this wff is negated and transformed to clause form, one obtains the three clauses Q(x), ~P(a), and ~Q(b), where a and b are Skolem constants. Clearly, Q(x) and ~Q(b) are contradictory, thus establishing that Q(x) implies P(x) | Q(y).

Example 1. Q(x) subsumes P(x) | Q(y), and the substitution is {y/x}. Q(x) does not s-subsume P(x) | Q(y), although P(x) does.

Example 2. P(x) | Q(y) subsumes P(a) | Q(g(a)) | ~R(x,b), and the substitution is {a/x,g(a)/y}.

Example 3. P(a,x) | P(y,b) subsumes P(a,b), and the substitution is {b/x,a/y}.

Example 4. P(x,y) subsumes P(a,u) | P(w,b). There are two possible substitutions: {a/x,u/y} and {w/x,b/y}.

Figure 4.6. Examples of subsumption.

## 4.10 Logical Soundness

As discussed in Section 2.4, a wff W logically follows from a set of wffs S if and only if every interpretation that satisfies wffs in S also satisfies W. Wffs can, of course, be replaced by clauses. For example, every interpretation that satisfies the clauses ~P(x) | Q(x) and P(a) also satisfies the clause Q(a), and thus Q(a) follows from ~P(x) | Q(x) and P(a). An inference procedure applied to a set of clauses S is **sound** if and only if every inferred clause W logically follows from S.

The procedures of binary resolution and binary factoring are sound. Consider binary resolution. Let C1 and C2 be two clauses, and let their resolvent C3 be formed by resolving on literal L1 in C1 and literal L2 in C2 using mgu μ. Assume literal L1 is uncomplemented. Then, C3 = [{C1 – L1} | {C2 – L2}]μ can be shown to follow logically from C1 and C2, as follows. Every interpretation that satisfies C1 and C2 also satisfies [C1]μ and [C2]μ, whereas no interpretation can satisfy both [L1]μ and [L2]μ, and thus every interpretation that satisfies C1 and C2 satisfies both [C1 –L1]μ and [C2 – L2]μ, which is C3. It is left for the reader to show that binary factoring is logically sound (see Exercise 4.6).

## 4.11 Base Clauses and Inferred Clauses

The set of **base clauses** are the clauses that constitute the axioms, the hypotheses, and the negated conclusion. When one attempts to find a proof, other **inferred clauses** are generated from the base clauses.

## Exercises for Chapter 4

Note that in the problems that follow, the symbols a, b, 2, 3, 4, and 5 represent constants.

4.1. For each of the following pairs of predicates, find their mgu and their most common instance — if one exists.

(a) P(x,y,z) and P(v,a,v)
(b) P(x,y,f(x)) and P(u,w,w)

(c) Q(u,u) and Q(x,f(z))
(d) P(a,x,f(g(y))) and P(z,f(z),f(u))
(e) R(x,f(x),z,f(z),u) and R(f(a),y,f(y),v,f(v))

4.2. Find all binary resolvents of each of the following pairs of clauses and indicate which are merge clauses.

(a) C1: P(a,x) | P(y,b) | Q(x) | R(y,b,x)
C2: ~P(u,v) | Q(v) | ~R(a,v,b)

(b) C3: P(x) | Q(x) | R(x)
C4: ~P(u) | ~Q(u) | ~R(a)

(c) C5: ~less(a,x) | ~less((f(a),x) | ~prime(x)
C6: less(h(u),u) | prime(u)

(d) C7: P(x,x) | ~P(x,h(x)) | ~Q(x,y)
C8: P(u,v) | Q(f(u),a)

(e) C9: P(x) | Q(y,h(y)) | Q(x,b)
C10: ~P(g(u,a)) | P(u) | ~Q(b,h(a))

4.3. Find all binary factors of the following clauses:

(a) C11: less(x,y) | less(y,x) | less(2,3) | ~less(5,4) | ~less(z,z)
(b) C12: P(a,x) | P(x,f(b)) | Q(x,a) | Q(f(y),y)
(c) C13: P(a,x) | P(x,f(a)) | P(a,f(a))

4.4. Suppose clause C1 has k literals and clause C2 has m literals. What is the maximum number of binary resolvents that these clauses can have? Give an example of two such clauses for k = m = 3.

4.5. Suppose clause C1 has k literals. What is the maximum number of binary factors that this clause can have? Give an example for k = 4.

4.6. Prove that binary factoring is a logically sound procedure.

4.7. Specify whether the first clause subsumes and/or s-subsumes the second clause for the following ten pairs of clauses.

(a) P(a) and P(a) | P(x)
(b) P(x) and P(a) | P(x)

(c) P(x) | Q(x) and P(a) | P(b) | Q(b)
(d) R(x,x) and R(y,b)
(e) R(f(x),a) and R(f(b),y)
(f) R(a,x) | R(y,b) and R(a,b)
(g) P(x,y) | Q(x) and P(x,a) | Q(a)
(h) P(x) | R(y,z) and P(a) | R(b,x) | T(w)
(i) R(x,y) | Q(y,x) and R(b,z) | Q(y,a)
(j) P(x,y) | Q(y) | R(y,x) and P(a,x) | Q(x) | R(x,y)

4.8. If one clause C1 subsumes another clause C2, it follows that C1 implies C2. First show that P(x) | R(y,z) subsumes P(a) | R(b,b) | T(w) and then using the technique of proof by contradiction, show that C1 implies C2.

4.9. If one clause, C1, subsumes another clause, C2, then C1 implies C2. However, if C1 implies C2, it is not necessarily true that C1 subsumes C2. Find two clauses C1 and C2 such that C1 implies C2 but C1 does not subsume C2.

# 5 Proving Theorems by Constructing Closed Semantic Trees

This chapter provides the theoretical foundations for proving theorems by constructing closed semantic trees. Section 5.1 introduces the Herbrand universe of a set of clauses, Section 5.2 introduces the Herbrand base of a set of clauses, and Section 5.3 introduces the concept of an interpretation on the Herbrand base. The use of a truth table to establish the unsatisfiability of a set of clauses is described in Section 5.4. The use of semantic trees for the same purpose is described in Section 5.5. Constructing noncanonical semantic trees is the subject of Section 5.6. In Chapter 7, HERBY, a program that constructs such noncanonical semantic trees, is described; HERBY uses a modified variation of Algorithm 5.1, which is described in Section 5.5.

## 5.1 The Herbrand Universe of a Set of Clauses

The Herbrand universe of a set of clauses S, denoted HU(S), is the set of all possible terms that can be formed from the functions and constants that appear in S. If there is no constant in S, one constant is arbitrarily placed in HU(S). That constant is named "a" in this text. The Herbrand universe is a finite set or, if not finite, a countably infinite set. If S has no functions, the set is finite. When the terms are listed recursively, they are said to be in **canonical order**.

**Example 5.1.** Consider HU1.THM in Figure 5.1. The canonical Herbrand universe of HU1.THM is the infinite set:

HU(HU1.THM) = {a,f(a),f(f(a)),f(f(f(a))), . . . }

Axioms
1 P(x)
2 ~P(a) | Q(x)

Negated conclusion
3 ~Q(f(x))

Figure 5.1. HU1.THM.

Axioms
1 P(h(x,y),x) | Q(x,y)
2 ~P(x,y) | Q(y,x)

Negated conclusion
3 ~Q(x,y)

Figure 5.2. HU2.THM.

**Example 5.2.** As a second example, consider HU2.THM in Figure 5.2. Because there is no constant in the theorem, the constant a is initially added to the Herbrand universe. The canonical Herbrand universe is then:

HU(HU2.THM) = {a, h(a,a), h(a,h(a,a)), h(h(a,a),a), h(h(a,a),h(a,a)),
h(a,h(a,h(a,a))), h(h(a,a),h(a,h(a,a))),
h(h(a,h(a,a)),a), h(h(a,h(a,a)),h(a,a)), . . . }

## 5.2 The Herbrand Base of a Set of Clauses

The Herbrand base of a set of clauses S, denoted HB(S), is defined as the set of all ground instances of the predicates of S where the arguments of the predicates are all possible combinations of the terms of the Herbrand universe of S. These instances are called **atoms**. The Herbrand base is a finite set, or, if not finite, a countably infinite set. When the atoms are listed recursively, they are said to be in **canonical order**.

**Example 5.3.** For HU1.THM, the canonical Herbrand base is:

HB(HU1.THM) = {P(a), Q(a), P(f(a)), Q(f(a)), P(f(f(a))), Q(f(f(a))), . . . }

**Example 5.4.** For HU2.THM, the canonical Herbrand base is:

HB(HU2.THM) = {P(a,a), Q(a,a), P(a,h(a,a)), Q(a,h(a,a)),
P(h(a,a),a), Q(h(a,a),a), . . . }

**Example 5.5.** For HU3.THM, the canonical Herbrand base is:

HB(HU3.THM) = {A, B, C}

## 5.3 An Interpretation on the Herbrand Base

An **interpretation** on the Herbrand base is an assignment of the value of TRUE or FALSE to each atom. If the number of atoms in the Herbrand base is finite — say there are N atoms — then the number of interpretations is $2^N$. Most theorems of interest have an infinite number of Herbrand base atoms and thus an uncountable number of interpretations.

An interpretation **fails to satisfy** (or **fails**) a clause if the atoms of some subset of Herbrand base atoms, when assigned values of TRUE or FALSE, conflict with the clause. The conflict can be established by showing that the atoms of the subset, with their assigned values, resolve with the literals of the clause to yield the NULL clause. Otherwise, an interpretation **satisfies** the clause. For example, for HU1.THM, any interpretation that assigns TRUE to atom P(a) and FALSE to atom Q(f(a)) fails clause 2 because P(a) resolves with the first literal of clause 2 (i.e., ~P(a)) to yield Q(x), and ~Q(f(a)) resolves with the resulting resolvent, Q(x), to yield the NULL clause.

An interpretation fails a set of clauses if it fails any clause of the set. Otherwise, it satisfies the set of clauses. A set of clauses is **satisfiable** if there is at least one interpretation that satisfies every clause. Otherwise, the set of clauses is **unsatisfiable**.

## 5.4 Establishing the Unsatisfiability of a Set of Clauses

There are a number of procedures that can be used to establish the unsatisfiability of the set of clauses that constitute a theorem's axioms, hypotheses, and negated conclusion. In this section, a truth table is used for this purpose and is illustrated by two examples. In the next section, semantic trees are introduced.

**Example 5.6.** Consider HU3.THM in Figure 5.3. The Herbrand base is {A,B,C}. There are eight interpretations, as shown in the truth table in Figure 5.4. Each interpretation fails the clause indicated, and thus the set of clauses that constitute HU3.THM is unsatisfiable. For example, the interpretation in which A is TRUE, B is TRUE, and C is TRUE, corresponding to the first row of the table, fails clause 1. The last of the eight interpretations, A is FALSE, B is FALSE, and C is FALSE, fails clauses 2 and 5. The six other intepretations also fail, and the clauses that each interpretation fails are given in the rightmost column.

**Example 5.7.** Consider again HU1.THM. The Herbrand base is given as {P(a), Q(a), P(f(a)), Q(f(a)), ... } in Section 5.2. There is an infinite number of

Axioms
1 ~A | ~C
2 A | C
3 ~A | C
4 A | ~C | B
5 B | C

Negated conclusion
6 A | ~C | ~B

Figure 5.3. HU3.THM.

| Interpretations on the Herbrand base | | | Unsatisfied clauses |
|---|---|---|---|
| A | B | C | |
| T | T | T | 1 |
| T | T | F | 3 |
| T | F | T | 1 |
| T | F | F | 3,5 |
| F | T | T | 6 |
| F | T | F | 2 |
| F | F | T | 4 |
| F | F | F | 2,5 |

Figure 5.4. The eight interpretations on the Herbrand base of HU3.THM and the clauses that each interpretation fails.

interpretations, as the truth table in Figure 5.5 shows. Each interpretation is shown to fail at least one clause. The first interpretation fails clause 3 because the fourth atom Q(f(a)) resolves with clause 3 to yield the NULL clause. The second interpretation fails clause 2 because the first and fourth atoms of the interpretation resolve with the literals of ~P(a) | Q(x) to yield the NULL clause. The third interpretation can be seen to fail clauses 1 and 3.

| Interpretations on the Herbrand base | | | | | Unsatisfied clauses |
|---|---|---|---|---|---|
| P(a) | Q(a) | P(f(a)) | Q(f(a)) | . . . | |
| T | T | T | T | | 3 |
| T | T | T | F | | 2 |
| T | T | F | T | | 1, 3 |
| T | T | F | F | | 2 |
| . . . | . . . | . . . | . . . | | |
| F | F | F | T | | 1, 3 |
| F | F | F | F | | 1 |

Figure 5.5. The intrepretations on the Herbrand base for HU3.THM.

In Example 5.7, each interpretation involves an assignment of a value of TRUE or FALSE to an infinite number of atoms. However, it is only necessary to list the first four atoms (in canonical order) to show that every interpretation fails at least one clause; that is, every **partial interpretation** on the first four atoms fails at least one clause of HU1.THM establishing the unsatisfiability of the set of clauses that constitute HU1.THM. In fact, every partial interpretation on just the third and fourth atoms — that is P(f(a)) and Q(f(a)) — fails at least one clause of the theorem, implying that four partial interpretations are sufficient to establish the unsatisfiability of the clauses of HU1.THM.

**Theorem 5.1 Herbrand's theorem (Version 1)**

If S is an unsatisfiable set of clauses, there is some k such that every partial interpretation over the first k atoms of the canonical Herbrand base fails S.

Instead of listing the interpretations in a truth table, they can be presented in an equivalent way, as the following section describes.

## 5.5 Semantic Trees

A **semantic tree** of depth D of a set of unsatisfiable clauses S is a binary tree (shown in this book as growing up). Each **node** is assigned clauses as described in the next paragraph. The **root node** is denoted by R. A node N is at depth d if it is d nodes away from the root along some path. Each node in the tree has either 0 or 2 children. In the first case, N is a **terminal node**; in the second case, N is a **nonterminal node**. **Branches** leading from nodes at depth d to nodes at depth d+1 are at depth d+1. R is thus at depth 0, and branches leading from R are at depth 1. Each branch is labeled with an atom or its negation. Assume the ordering of atoms is $hb_1$, $hb_2$, ..., $hb_d$ .... Then, each left branch at depth d is labeled with $hb_d$ whereas each right branch is labeled with ~$hb_d$. All nodes at depth D are terminal nodes. A node at depth d less than D is also a terminal node if it is a failure node, as defined shortly.

Each node N is assigned a set of clauses, denoted K(N), as follows:

1. The root node R is assigned the given set of clauses S (i.e., K(R) = S).
2. For any other node N with the nodes on the path to it labeled R, $N_a$, ..., $N_t$ and with the branch leading immediately to it labeled with atom hb or its negation, ~hb, K(N) is assigned all resolvents of hb or its negation with all clauses in the sets K(R), K($N_a$), ..., K($N_t$) and with the clauses so generated. However, do not add to K(N) a resolvent if it already appears in K(R), K($N_a$), ..., K($N_t$) — or in K(N).

A node N is a **failure node** (and thus a terminal node) if the NULL clause is assigned to it. In this case, the partial interpretation specified by the set of atoms on the path from R to N fails at least one clause in S.

A **canonical semantic tree** of depth D is a semantic tree of depth D in which each left branch at depth $d \leq D$ is labeled with the dth atom and each right branch at depth $d \leq D$ is labeled with the complement of the dth atom of the canonical Herbrand base. A **noncanonical semantic tree** has branches labeled with atoms from some partial interpretation.

**Theorem 5.2. Herbrand's theorem (Version 2)**
If S is an unsatisfiable set of clauses, there exists some k such that every path in a canonical semantic tree for S beginning at the root and of length at most k leads to a failure node. The semantic tree is said to be **closed** in this case.

We thus have a way to prove a theorem. We simply show that a closed canonical semantic tree exists for the set of clauses that constitute the axioms and negated conclusion. Of course, if any closed semantic tree can be constructed, canonical or noncanonical, a proof has been established.

**Algorithm 5.1. Proving a theorem by constructing a closed semantic tree**

1. Set the depth D of the semantic tree under construction to 0.
2. Carry out a depth-first construction of a semantic tree of depth D.
   - 2a. If all nodes at depth D or less are found to be failure nodes, the construction is complete and the algorithm terminates.
   - 2b. If and whenever a node M at depth D is found not to be a failure node, increase D by one, choose some atom from the Herbrand base, and thereafter label all branches at depth D with this atom or its negation. Do not choose an atom that has been previously selected. Continue the construction from M.

Choosing the atoms in Step 2b in canonical order is one possible approach, but it is not particularly effective. In Chapter 7, the atom selection heuristics of HERBY are described. It should be pointed out that Algorithm 5.1 is both sound and complete.

**Example 5.8.** Let S be the unsatisfiable set of clauses that constitute HU3.THM, as shown in Figure 5.3. Figure 5.6 shows the semantic tree constructed by Algorithm 5.1. The order in which atoms were selected was $hb_1 = C$, $hb_2 = A$, and $hb_3 = B$.

Initially, an attempt is made to construct a closed semantic tree of depth 0. The root node is assigned the set of clauses S (i.e., K(R) = S). Because the

NULL clause is not in K(R), the depth of the semantic tree under construction is increased to 1. Then the branch from R labeled with atom C is constructed and we arrive at node N1. The set of clauses in K(N1) is found by resolving C with the clauses in K(R) and is:

K(N1) = {7: (1b) ~A, 8: (4b) A I B, 9: (5b) A I ~B}

The notation describing clause 7 says that it is formed by resolving the atom on the branch leading to N1 — that is atom C — with the second, or "b" literal of clause 1 yielding the clause ~A. Because N1 is at depth 1 and is not a failure node, D is increased to 2 and the construction continues. The branch labeled with atom A is constructed, and node N2 is arrived at next. The set of clauses in K(N2) is found by resolving A with the clauses in K(R) and K(N1) and is:

K(N2) = {10: (7a) Ø, etc.}

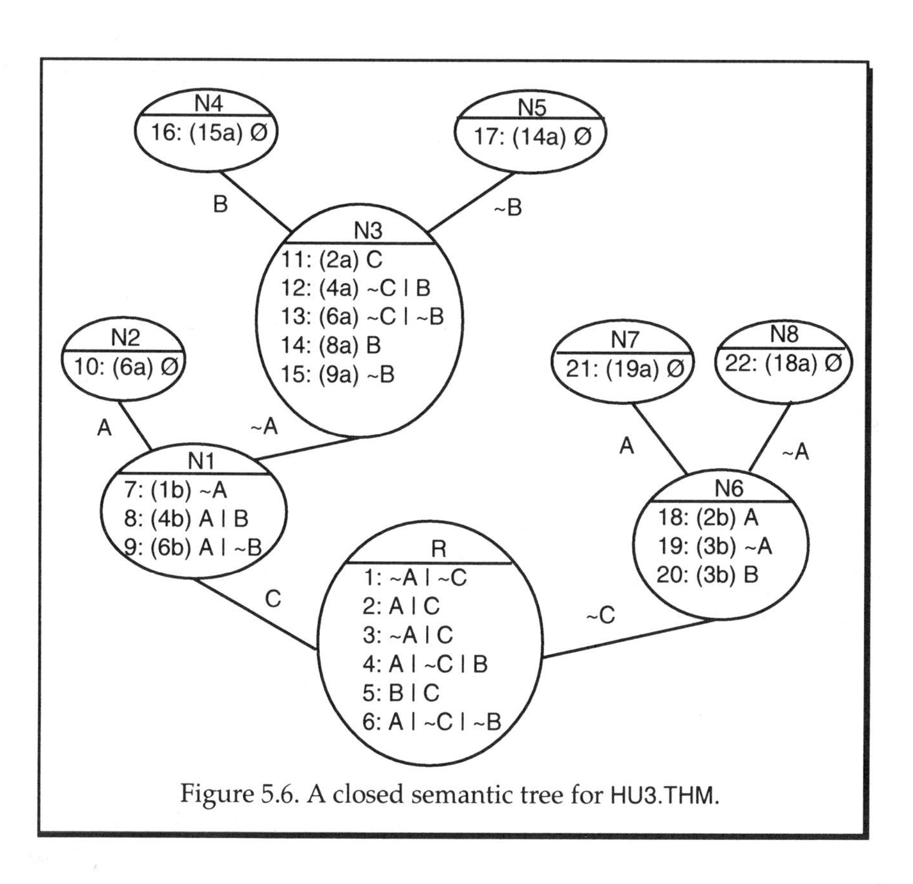


Figure 5.6. A closed semantic tree for HU3.THM.

Once the NULL clause is generated at a node, it is unnecessary to generate other clauses at that node, and thus only clause 10 is shown. Clause 10 is generated by resolving atom A with the first literal of clause 7. Node N2 is a failure node and thus a terminal node. The construction continues by constructing the branch labeled with ~A and we arrive at node N3. The set of clauses in K(N3) is found by resolving ~A with the clauses in K(R) and K(N1) and is:

K(N3) = {11: (2a) C, 12: (4a) ~C | B, 13: (5a) ~C | ~B, 14: (7a) B, 15: (8a) ~B}

Because N3 is at depth 2 and is not a failure node, D is increased to 3, and the construction continues. The branch labeled with atom B is constructed and we arrive at node N4. The set of clauses in K(N4) is found by resolving B with the clauses in K(R), K(N1), and K(N3) and is:

K(N4) = {16: (15a) Ø, etc.}

N4 is a failure node and thus a terminal node. The construction continues by labeling the right branch at N3 with ~B, and we arrive at node N5. The reader can verify that N5 is a terminal node. The construction then backtracks to the root and constructs the right branch of R, labeling it with ~C and leading to node N6, where:

K(N6) = {18: (2b) A, 19: (3b) ~A, 20: (5b) B}

Lastly, the children of N6 — that is, nodes N7 and N8 — can be seen to be terminal nodes, and thus a closed semantic tree is constructed.

Proving theorems by constructing closed semantic trees was the subject of Almulla's doctoral thesis at McGill University. He measured the effectiveness of this approach on the theorems in the Stickel test set and found that closed canonical semantic trees could be constructed for 29 of the 84 theorems. Many theorems would have required vast amounts of time to solve. Nevertheless, it is interesting that even 29 theorems could be proved using this simple approach. In Chapter 7, heuristics used by HERBY that attempt to select atoms in a more effective order are described.

## 5.6 Noncanonical Semantic Trees

If a partial interpretation on atoms $hb_1$, $hb_2$, ..., $hb_k$ yields a closed semantic tree, then any semantic tree constructed by labeling the branches in any of the other k! – 1 orderings of these atoms also yields a closed semantic tree. In such cases, **noncanonical semantic trees** are constructed. Moreover, a noncanonical semantic tree could have different atoms labelling the branches of the same level in the tree. This is illustrated by HU4.THM in Figure 5.7 and the closed semantic tree in Figure 5.8. Note that at the second level in the semantic tree two different atoms label the branches.

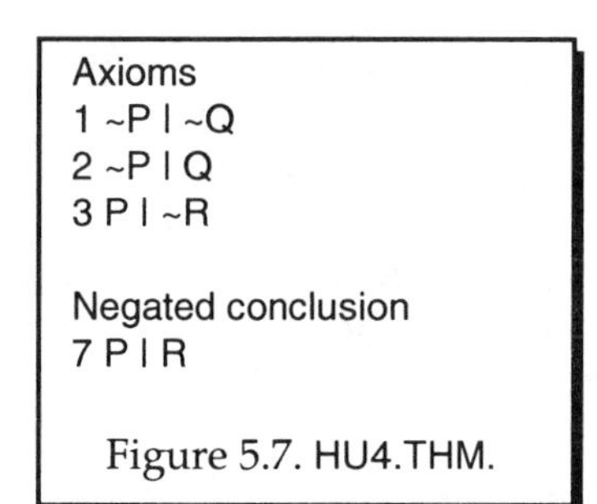


Figure 5.7. HU4.THM.

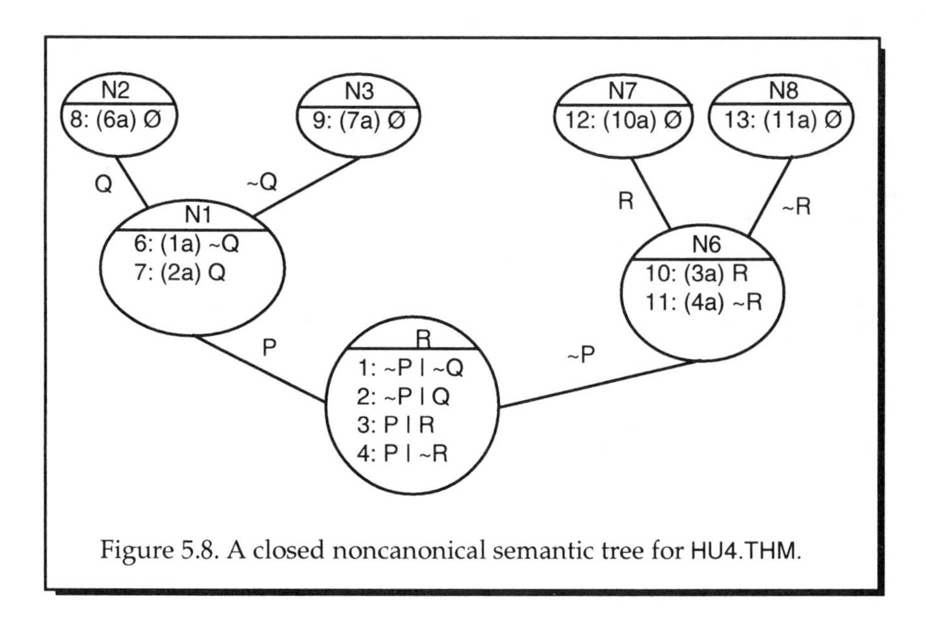


Figure 5.8. A closed noncanonical semantic tree for HU4.THM.

## Exercises for Chapter 5

5.1. List the first 20 terms of the canonical Herbrand universe and the first 20 atoms of the canonical Herbrand base for HU2.THM (see Figure 5.2).

5.2. How many terms are there in the Herbrand universe of S27WOS3A.THM (see Figure 5.12), and how many atoms are there in the Herbrand base?

5.3. Construct a truth table similar to the one in Figure 5.4 to prove EX2.THM (see Figure 5.9).

5.4. Construct a closed canonical semantic tree for EX2.THM (see Figure 5.9).

5.5. Construct a closed canonical semantic tree for FACT.THM (see Figure 5.10).

5.6. Construct a closed canonical semantic tree for EX3.THM (see Figure 5.11).

5.7. Construct a closed (noncanonical) semantic tree for S27WOS3A.THM having the minimum number of atoms (see Figure 5.12).

```
Axioms
1 ~P | ~Q | ~R     2 ~P | ~Q | R
3 ~P | Q | ~R      4 ~P | Q | R
5 P | ~Q | ~R      6 P | ~Q | R

Negated conclusion
7 P | Q
```

Figure 5.9. EX2.THM.

```
Axiom
1 P(x) | P(y)

Negated conclusion
2 ~P(x) | ~P(y)
```

Figure 5.10. FACT.THM.

```
Axioms
1 p(a)
2 ~p(x) | p(g(x))

Negated conclusion
3 ~p(g(g(g(x))))
```

Figure 5.11. EX3.THM.

```
Axioms
1 p(e,x,x)
9 ~p(x,y,z) | ~p(x,y,u) | Equal(z,u)
19 p(x,a,x)

Negated conclusion
20 ~Equal(e,a)
```

Figure 5.12. S27WOS3A.THM.

# 6 Resolution–Refutation Proofs

This chapter establishes the theoretical foundations of resolution–refutation theorem proving as carried out by THEO. A **resolution–refutation proof** is a proof in which some sequence of inferences performed on a theorem's base clauses and on resulting inferences derives the NULL clause. Inferences generated by THEO are restricted to binary resolution and binary factoring.

In the previous chapter, it was shown that a closed semantic tree can be constructed for the clauses that constitute a theorem's base clauses. In this chapter, Algorithm 6.1 is presented and shown to transform such a closed semantic tree to a resolution–refutation proof. A consequence of this algorithm is that there exists a procedure for obtaining a resolution–refutation proof for every theorem. The procedure for constructing a closed semantic tree and then using the tree to obtain a resolution–refutation proof is not particularly practical. More effective procedures are presented in Chapter 9.

Section 6.1 presents four examples of resolution–refutation proofs, although sidestepping the issue of how they were actually obtained. A procedure for finding such proofs is called a **resolution–refutation search procedure**. Many search strategies can be used. The ones used by THEO are described in Chapter 9. A resolution–refutation proof can be illustrated by a **resolution–refutation proof graph**, and this is done for the four examples.

In Section 6.2, the terms depth and length of a resolution–refutation proof are defined. In Section 6.3, it is shown that if a proof of a theorem exists, a resolution–refutation proof can always be found. Sections 6.4 and 6.5 show that certain restrictions can be made on the structure of a proof. These restrictions often reduce the size of the search space without eliminating inferences required to obtain a proof.

The material in Sections 6.4 and 6.5 assumes all clauses are ground clauses. Section 6.6 extends the material to the more general case of predicate calculus. Section 6.7 is concerned with handling binary factoring.

## 6.1 Examples of Resolution–Refutation Proofs

**Example 6.1.** Q07B0.THM, shown in Figure 6.1, uses three axioms (clauses 4, 20, and 24) from the set given as wffs in Section 2.6. Clauses 39 and 40 are the negated conclusion, which, before negated, says: for all x,y,z,u,v, if x = Ext(y,z,u,v) — extending segment y–z by segment u–v — then z is between y and x, or as a wff: @x@y@z@u@v: Equal(x,Ext(y,z,u,v)) => B(y,z,x).

Four binary resolutions lead to the NULL clause, as shown in Figure 6.2. Clauses in proofs are always numbered to follow the last clause of the negated conclusion so the proof arbitrarily begins with clause 41. The resolution–refutation proof graph is shown in Figure 6.3. Each node corresponds to a base clause or to an inference; branches leading to a node originate at the input clauses of the node.

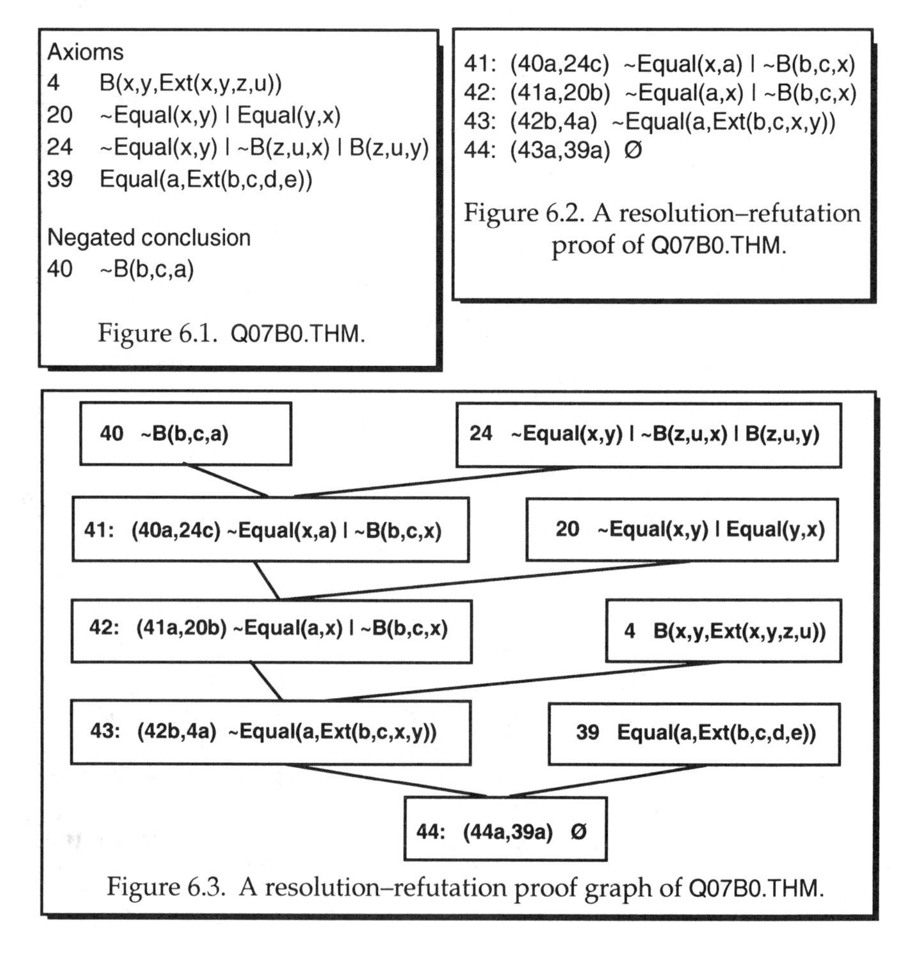


Figure 6.1. Q07B0.THM.



Figure 6.2. A resolution–refutation proof of Q07B0.THM.



Figure 6.3. A resolution–refutation proof graph of Q07B0.THM.

**Example 6.2.** Consider A.THM in Figure 6.4. It has two axioms and a negated conclusion. Two proofs are shown in Figure 6.5; the corresponding resolution–refutation proof graphs are shown in Figure 6.6. This example illustrates that a theorem can have more than one proof.

**Example 6.3.** STARK017.THM in Figure 6.7 says "there exists an infinite number of primes." The first ten clauses are axioms, the eleventh the "if" side of the negated conclusion, and the twelfth the "then" side. The function F(x) is x! + 1 and H(x) is a Skolem function introduced when the original wffs were converted to clauses. Clauses 8, 9, and 10 come from a wff that said: for all

Axioms
1 P(x) | ~Q(x)
2 Q(a) | Q(b)

Negated conclusion
3 ~P(x)

Figure 6.4. A.THM.

Proof 1:
4: (1a,3a) ~Q(x)
5: (4a,2a) Q(b)
6: (4a,5a) Ø

Proof 2:
4: (1b,2a) P(a) | Q(b)
5: (4a,3a) Q(b)
6: (5a,1b) P(b)
7: (6a,3a) Ø

Figure 6.5. Two resolution–refutation proofs of A.THM.

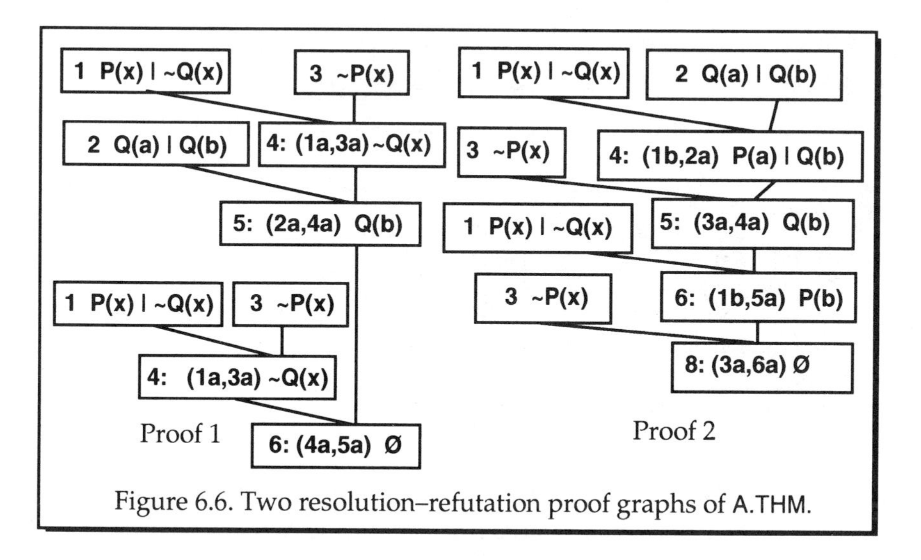

Figure 6.6. Two resolution–refutation proof graphs of A.THM.

x, if x is not prime, there exists a y such that y divides x, y is prime, and y is less than x. Clauses 11 and 12 come from negating a wff that says: for all x, there exists a y such that if x is prime, then y is prime and x is less than y and F(x) is not less than y. One resolution–refutation proof is given in Figure 6.8; the corresponding resolution–refutation proof graph appears in Figure 6.9.

```
Axioms
1    ~Less(x,x)                       ; x ≮ x
2    ~Less(x,y) | ~Less(y,x)          ; if x < y, then y ≮ x
3    Divides(x,x)                     ; x divides x
4    ~Divides(x,y) | ~Divides(y,z)    ; if x divides y, y divides z,
            | Divides(x,z)              then x divides z
5    ~Less(x,y) | ~Divides(y,x)       ; if x <  y,
                                        then y does not divide x
6    Less(x,y) | ~Divides(y,F(x))     ; if y divides F(x), then x < y
7    Less(x,F(x))                     ; x < F(x)
8    Prime(x) | Divides(H(x),x)       ; if x is not prime, then there is a y
                                        such that y divides x and
9    Prime(x) | Prime(H(x))           ; y is prime and y is less than x.

10   Prime(x) | Less(H(x),x)
11   Prime(a)                         ; Hypothesis: If a is prime

Negated conclusion
12   ~Less(a,x) | Less(F(a),x)        ; then x is not prime or a is not less than
          | ~Prime(x)                   x or F(a) is less than x
```

Figure 6.7. STARK017.THM.

```
13:  (12c,9a)    ~Less(a,x) | Less(F(a),x) | Prime(H(x))
14:  (12c,8a)    ~Less(a,x) | Less(F(a),x) | Divides(H(x),x)
15:  (13c,12c)   ~Less(a,x) | Less(F(a),x) | ~Less(a,H(x)) | Less(F(a),H(x))
16:  (15c,6a)    ~Less(a,x) | Less(F(a),x) | Less(F(a),H(x)) | ~Divides(H(x),F(a))
17:  (16c,5a)    ~Less(a,x) | Less(F(a),x) | ~Divides(H(x),F(a))
18:  (17c,14c)   ~Less(a,F(a)) | Less(F(a),F(a))
19:  (18a,7a)    Less(F(a),F(a))
20:  (19a,1a)    Ø
```

Figure 6.8. A resolution–refutation proof of STARK017.THM.

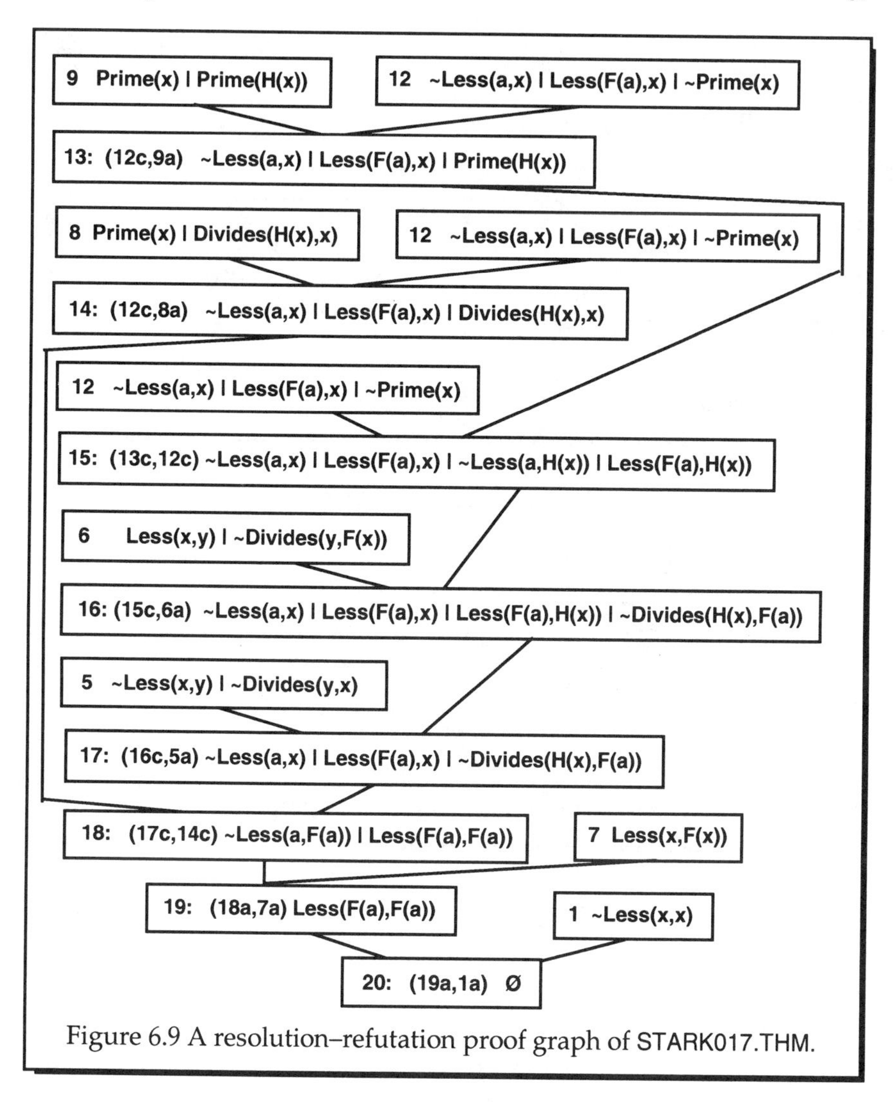

Figure 6.9 A resolution–refutation proof graph of STARK017.THM.

**Example 6.4.** Consider FACT.THM, shown in Figure 5.10 and repeated here in Figure 6.10. The binary resolvent 3: (1a,2a) P(x) | ~P(y) is the only clause that can be produced by resolving together clauses 1 and 2. Further, clause 3 does not produce any new clauses when resolved with either clause 1 or clause 2. Thus, this example establishes that not all proofs can be obtained using only binary resolution. However, if a binary factor of clause 1 is produced, namely 3: (1ab) P(x), it becomes possible to generate the NULL clause as shown in Figure 6.11. The corresponding resolution–refutation proof graph is shown in Figure 6.12.

Axiom
1 P(x) | P(y)

Negated conclusion
2 ~P(x) | ~P(y)

Figure 6.10. FACT.THM.

3: (1ab) P(x)
4: (3a,2a) ~P(x)
5: (3a,4a) Ø

Figure 6.11. A resolution–refutation proof of FACT.THM.

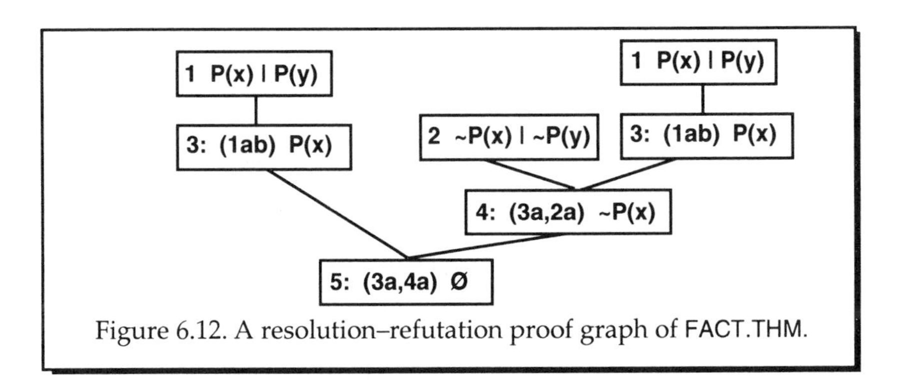


Figure 6.12. A resolution–refutation proof graph of FACT.THM.

## 6.2 The Depth and Length of Resolution–Refutation Proofs

The **depth** of a resolution–refutation proof is the number of inferences on the longest path in the resolution–refutation proof graph from the root to a base clause. The **length** of a resolution–refutation proof is the total number of inferences in the proof.

## 6.3 Obtaining a Resolution–Refutation Proof from a Closed Semantic Tree

In Section 6.1, resolution–refutation proofs of four theorems are presented. Theorem 6.1 shows that every theorem has a resolution–refutation proof.

**Theorem 6.1. Every theorem has a resolution–refutation proof**
Proof: This is established by beginning with a closed semantic tree — there is a closed semantic tree for every theorem — and then deriving a resolution–refutation proof from the tree by using Algorithm 6.1. (Better ways exist for finding resolution–refutation proofs, as shown in Chapter 9, but the purpose here is only to show that every theorem has such a proof.)

Assume now that a closed semantic tree has been constructed for some theorem T. Based on the semantic tree, the algorithm for obtaining a resolution–refutation proof repeatedly (1) finds two failure nodes that are siblings, (2) resolves together the clauses that fail at these nodes to generate additional clauses of the resolution–refutation proof, and then (3) generates a new closed semantic tree based on the set of base clauses and those clauses added in Step 2. When the NULL clause is generated, the algorithm terminates with a resolution–refutation proof.

**Algorithm 6.1. Obtaining a resolution–refutation proof**
Step 1. Select two failure nodes N1 and N2 that are siblings of some node N in the semantic tree and that fail because of clauses C1 and C2, respectively. If there is no such pair, terminate the algorithm. Let Ca and Cb be copies of the clauses C1 and C2, respectively. Let A and ~A be the atoms on the branches leading to N1 and N2, respectively.

Step 1a. (Factor Ca) If two literals of Ca were resolved away by A, form a factor of Ca by factoring on these two literals. Replace Ca with this factor. Add the new Ca to the resolution–refutation proof.

(A semantic tree constructed with the same set of atoms and for the set of clauses in T and those added thus far to the proof before adding Ca fails at exactly the same nodes as does a semantic tree constructed after adding Ca.)

Step 1b. (Try factoring Ca again) If there remain two literals of Ca that were resolved away by A, return to Step 1a. Otherwise, go to Step 1c.

Step 1c. (Factor Cb) If two literals of Cb were resolved away by atom ~A, form a factor of Cb by factoring on these two literals. Replace Cb with this factor. Add the new Cb to the resolution–refutation proof.

Step 1d. (Try factoring Cb again) If there remain two literals of Cb that were resolved away by atom ~A, return to Step 1c. Otherwise, proceed to Step 1e.

Step 1e. (Resolve Ca,Cb) Form the resolvent of Ca and Cb by resolving on the literals of Ca and Cb that were resolved away by A and ~A, respectively; call the resolvent Cc. Add Cc to the resolution-refutation proof.

(Note that a semantic tree constructed with the same set of atoms and for the set of clauses in T and those added thus far to the proof including Cc is a subtree of the semantic tree constructed before adding Cc.)

Step 1f. (Factor Cc) If two literals of Cc were resolved away by the same atom, form the factor of Cc that results by factoring on these two literals and replace Cc with this factor. Add the new Cc to the resolution–refutation proof.

Step 1g. (Try factoring Cc again) If there remain two literals in Cc that were resolved away by the same atom, return to Step 1f. Otherwise, go to Step 2.

Step 2. Add Cc to the set of base clauses. If Cc is the NULL clause, the proof is complete and the algorithm terminates.

Step 3. Construct a new semantic tree for the enlarged set of base clauses. This new semantic tree will have at least one less node than the previous semantic tree, failing at all nodes that its predecessor did and failing at node N as well due to the new clause just added in Step 2. Return to Step 1.

When the algorithm terminates, the extended set of base clauses will contain the clauses that constitute a resolution–refutation proof. This algorithm is a proof of the completeness of binary resolution and binary factoring for resolution–refutation proofs.

**Example 6.5.** Figure 6.13 is the canonical semantic tree for HU1.THM (see Figure 5.1). There are only two failure nodes in the tree that are siblings. Resolving together the clauses that fail at these nodes on the appropriate literals gives 4: (3a,2b) ~P(a), as shown in Figure 6.13. The semantic tree for the new set of clauses is shown in Figure 6.14. The NULL clause falls out: 5: (4a,1a) Ø. The resolution–refutation proof is presented in Figure 6.15.

**Example 6.6.** Figure 6.16 is a noncanonical semantic tree for HU2.THM (see Figure 5.2). There are several ways to choose the first two failure nodes that are siblings. Arbitrarily choose the top leftmost pair and form clause 4: (2a,1a) Q(x,y) | Q(x,h(x,y)), as shown in Figure 6.16. Then, construct the semantic tree for this modified set of clauses containing the three base clauses and clause 4, as shown in Figure 6.17. Again, there are several choices for the second resolvent. Arbitrarily form clause 5: (3a,4b) Q(x,y), as shown in Figure 6.17. Then, again, construct the semantic tree for the set of

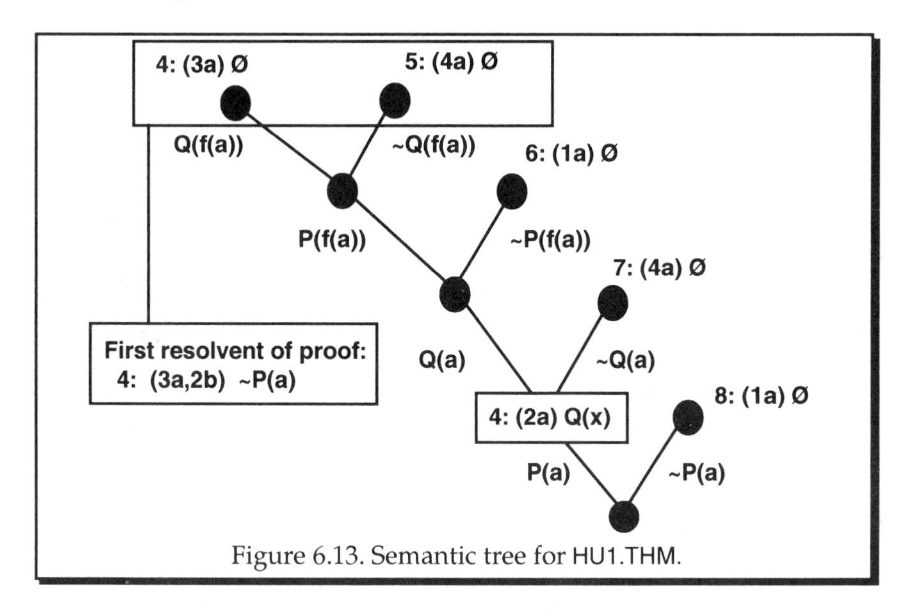


Figure 6.13. Semantic tree for HU1.THM.

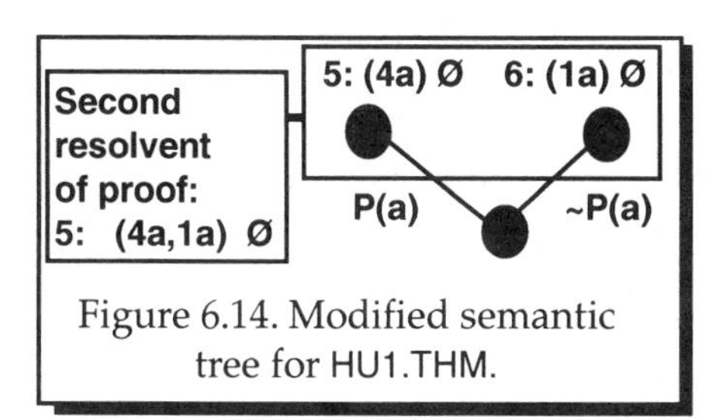


Figure 6.14. Modified semantic tree for HU1.THM.

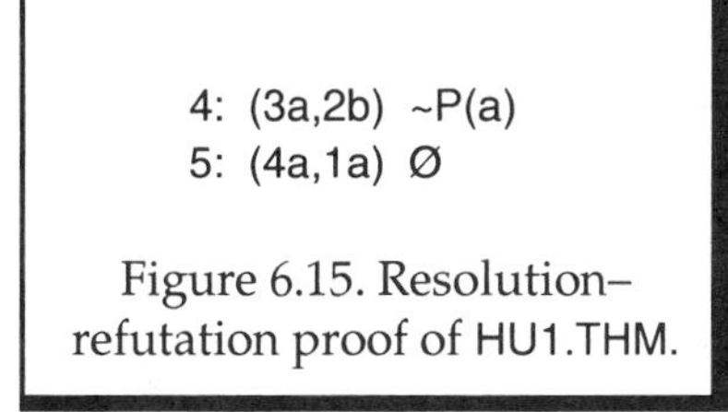
4: (3a,2b) ~P(a)
5: (4a,1a) Ø

Figure 6.15. Resolution–refutation proof of HU1.THM.

clauses containing the original base clauses and clauses 4 and 5 as shown in Figure 6.18, and finally, form clause 6: (3a,5a) Ø. The resolution–refutation proof is shown in Figure 6.19.

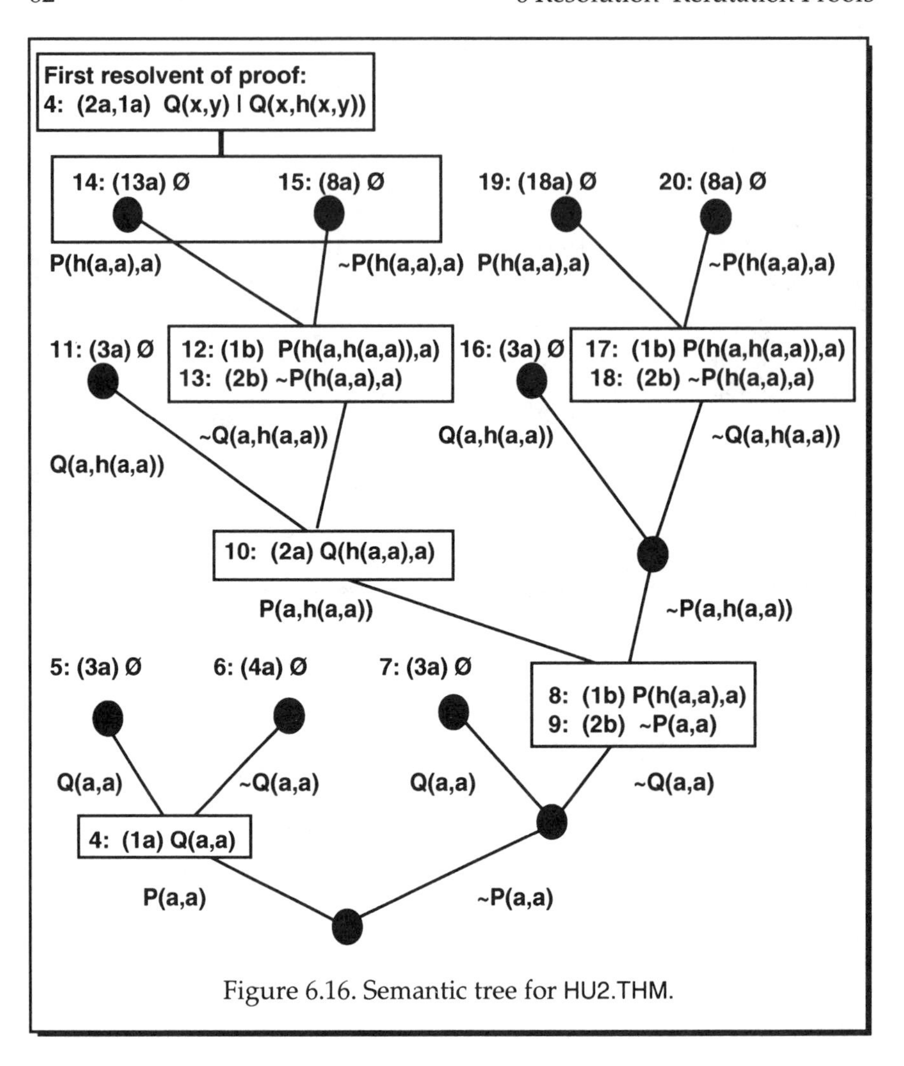


Figure 6.16. Semantic tree for HU2.THM.

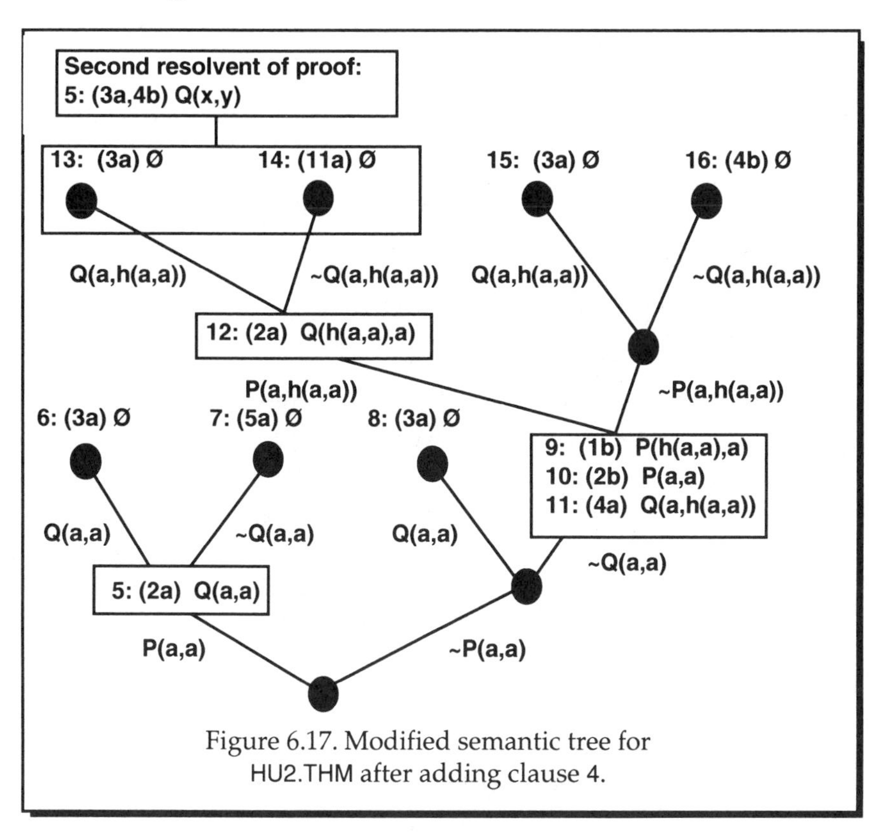


Figure 6.17. Modified semantic tree for HU2.THM after adding clause 4.

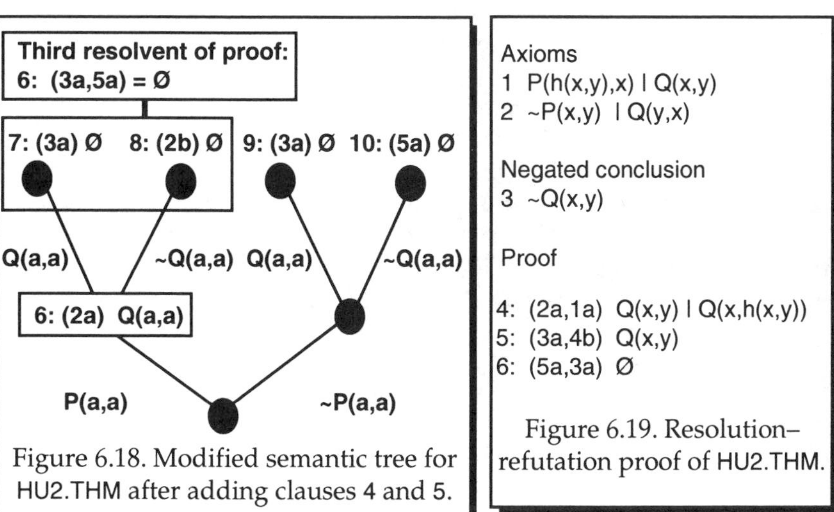


Figure 6.18. Modified semantic tree for HU2.THM after adding clauses 4 and 5.

Axioms
1 P(h(x,y),x) | Q(x,y)
2 ~P(x,y) | Q(y,x)

Negated conclusion
3 ~Q(x,y)

Proof

4: (2a,1a) Q(x,y) | Q(x,h(x,y))
5: (3a,4b) Q(x,y)
6: (5a,3a) Ø

Figure 6.19. Resolution–refutation proof of HU2.THM.

## 6.4 Linear Proofs

The material in this and the following section assumes all clauses are ground clauses and factoring does not occur. In Section 6.6, the "lifting lemma" is presented. This lemma justifies extending this material to the more general case of first-order predicate calculus. When all clauses are ground clauses, and if no base clause has two identical literals, factoring does not occur; all inferences are resolutions, some of which may be merge clauses. In Section 6.7, factoring is considered.

A proof of a theorem is **linear** if there is a single path of resolvents in the resolution–refutation proof graph beginning with some base clause and ending at the NULL clause, and furthermore, no resolvent is off this path. Let $R_1, R_2, ..., R_n = \emptyset$ denote the resolvents on this path of **length n**. For all i, $R_i$ is the **parent** of $R_{i+1}$ and an **ancestor** of $R_{i+1}, R_{i+2}, ..., R_n$. For a linear proof, resolvent $R_1$ has two base clauses as inputs, whereas for $i > 1$, $R_i$ has $R_{i-1}$ as one input and either a base clause or an $R_k$ for some $k < i - 1$ as the other. Similarly, a derivation of any clause, not alone the NULL clause, from another set of given clauses is call linear if all resolvents in the derivation are on a single path. A merge-free derivation is a derivation in which no resolvent is a merge clause.

The reader might note that except for the proof of STARK017.THM shown in Figure 6.9, the resolution–refutation proofs presented earlier in this chapter are either linear or can be easily transformed to linear. The proof of Q07B0.THM in Figure 6.3 is linear, as is Proof 2 of A.THM in Figure 6.6. The proof of FACT.THM, shown in Figure 6.12, can easily be transformed to linear form. Every theorem has a linear proof, as described in Theorem 6.2

**Theorem 6.2. Every theorem has a linear proof**
Proof: It was established in Section 6.3 that a closed semantic tree exists for every theorem. From the closed semantic tree, it was shown how to construct a resolution-refutation proof, which can be drawn as a resolution–refutation proof graph. It is easy to transform the graph to a **resolution–refutation proof tree**. The following algorithm shows how to transform the tree to a linear resolution–refutation proof, or more simply, to a linear proof.

**Algorithm 6.2. Obtaining a linear proof**
Begin with a resolution–refutation proof tree with M resolvents in the tree.

(1) Choose any path in the resolution–refutation proof tree leading from some base clause, say CB, to the root clause, $\emptyset$. Call this the **principal path**. Let the clauses on this path be CB, $R_1, R_2, ..., R_k, ..., R_m = \emptyset$. If $m = M$, the proof

is already linear; quit the algorithm. Otherwise, $m < M$; go to Step 2.

(2) Set $k = 2$.

(3) Consider the subtree rooted at $R_k$. There are two cases to consider as shown in Figure 6.20. In Case 1, where $R_{k-1}$ is the root of a linear derivation and C is a base clause or resolvent $R_j$ where $j < k-1$, there is no need to modify the derivation rooted at $R_k$; go to Step 4. In Case 2, where $R_{k-1}$ is the root of a linear derivation and C is the resolvent of two clauses, $C_a$ and $C_b$, there are two subcases to consider: Case 2a, where C is not a merge clause in some literal a; and Case 2b, where C is a merge clause in that literal. This is shown in Figure 6.21.

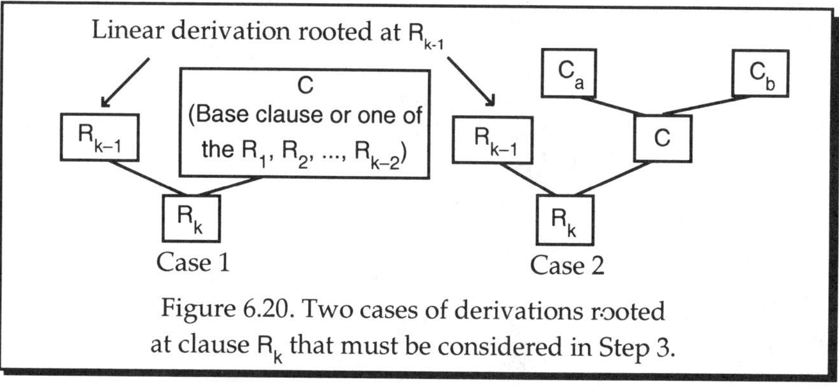


Figure 6.20. Two cases of derivations rooted at clause $R_k$ that must be considered in Step 3.

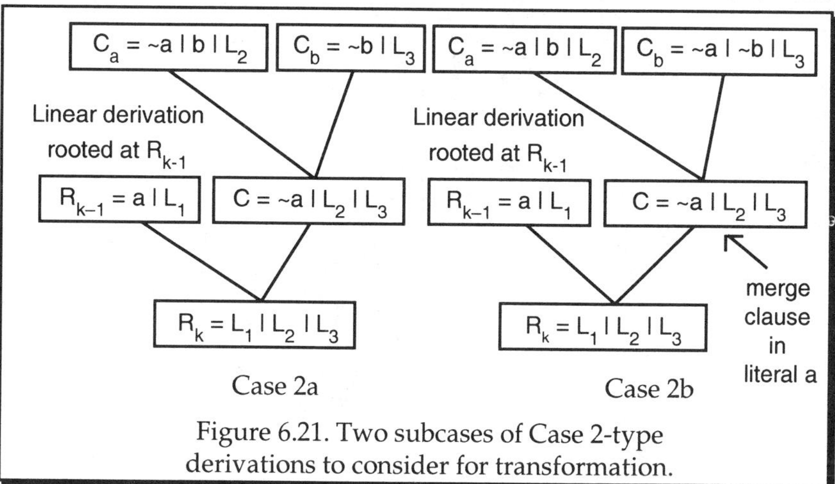


Figure 6.21. Two subcases of Case 2-type derivations to consider for transformation.

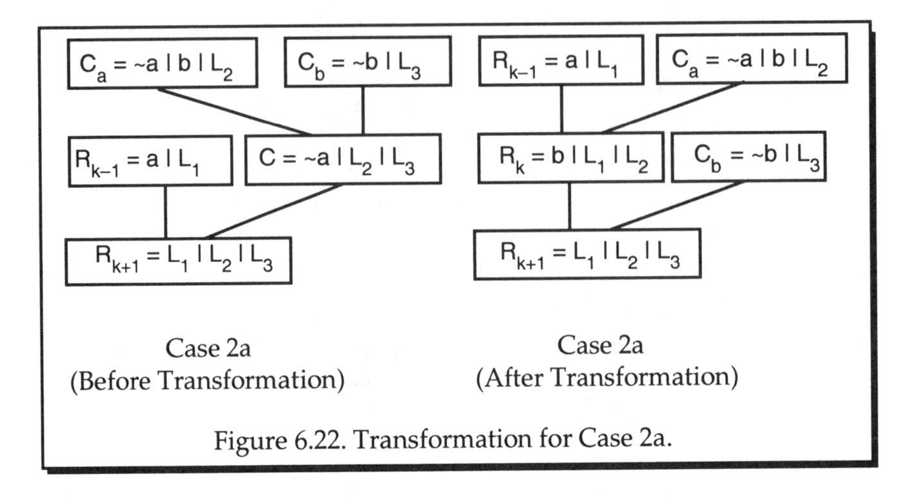


Figure 6.22. Transformation for Case 2a.

In Case 2a where C is not a merge clause in literal ~a, (and $L_1$, $L_2$ and $L_3$ are each a disjunction of literals) the proof can be transformed as shown in Figure 6.22. One new clause is added to the principal path while the number of clauses off the principal path is reduced by one. Rename clauses on the principal path CB, $R_1$, $R_2$, ..., $R_{n+1}$ = Ø, set m = m + 1, and repeat Step 3.

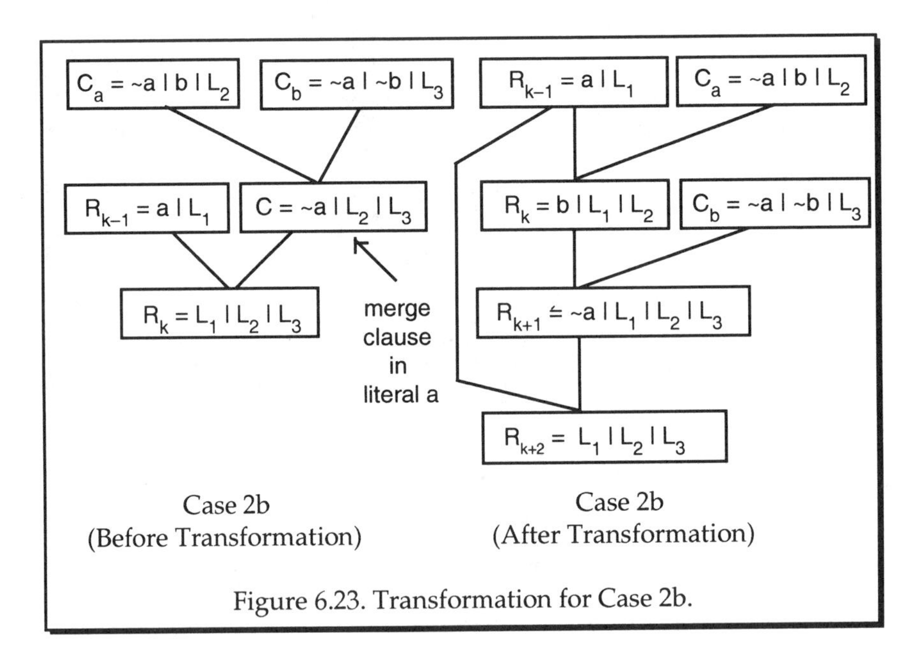


Figure 6.23. Transformation for Case 2b.

In Case 2b, where C is a merge clause in literal ~a, the proof can be transformed as shown in Figure 6.23. In this case, two new clauses are added to the principal path, and the number of clauses not on the principal path is reduced by one. Rename clauses on the principal path CB, $R_1$, $R_2$, ..., $R_{m+1}$, $R_{m+2}$ = Ø, set m = m + 2, and return to Step 3.

(4) Set k = k + 1. If k > m, quit; otherwise go to Step 3.

Each transformation reduces the number of clauses off the principal path by one, and thus the algorithm must terminate. It is left for the reader in Exercise 6.6 to establish an upper bound on the final length of the principal path.

**Example 6.7.** Consider LINEAR.THM in Figure 6.24. A resolution–refutation proof tree is shown in Figure 6.25. Let the principal path arbitrarily be the one leading to base clause 4, as shown by the heavy line. There are four resolvents off the principal path, and thus four transformations are necessary to obtain a linear form. They are shown in Figures 6.26–6.29. The number in each circle indicates the order in which the resolvents off the principal path are eliminated.

Axioms
1 A | B | C | D
2 A | B | C |~D
3 A | B | ~C
4 A | ~B | C
5 A | ~B | ~C
6 ~A | B
7 ~A | ~B | C

Negated conclusion
8 ~A | ~B | ~C

Figure 6.24. LINEAR.THM.

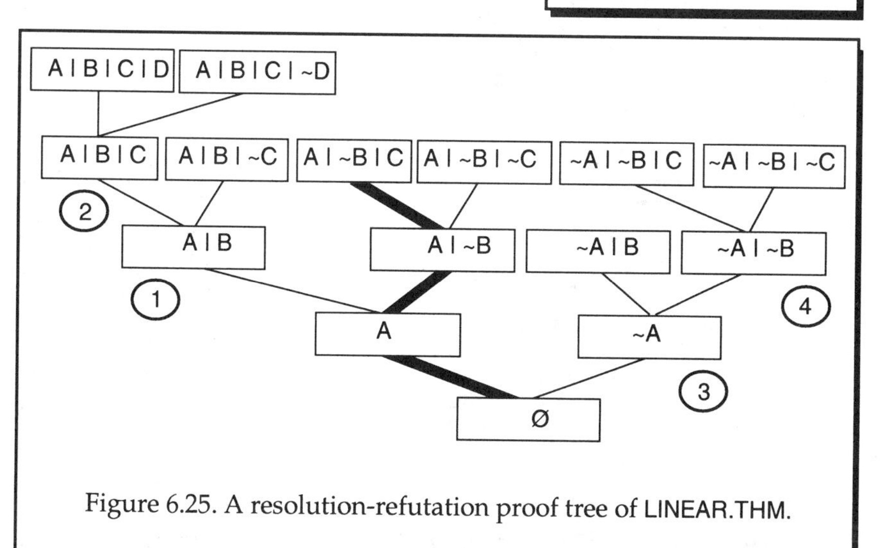


Figure 6.25. A resolution-refutation proof tree of LINEAR.THM.

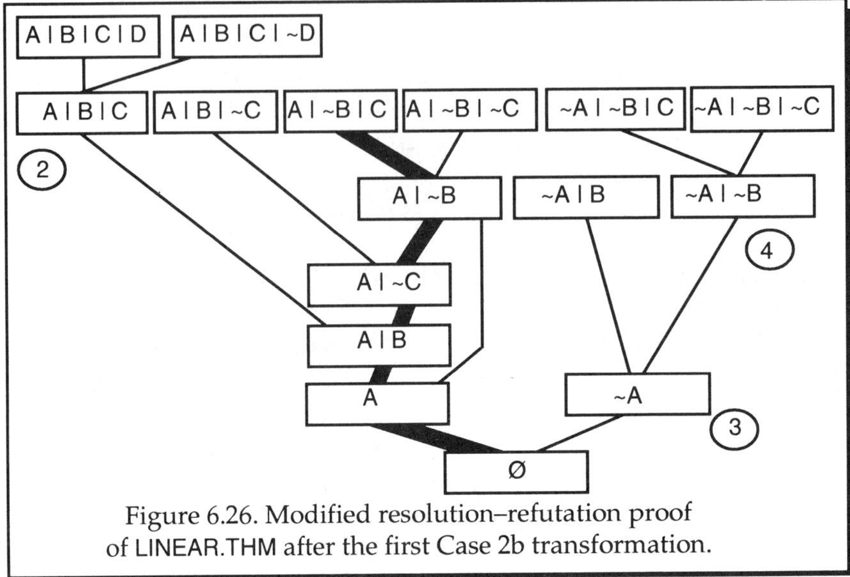


Figure 6.26. Modified resolution–refutation proof of LINEAR.THM after the first Case 2b transformation.

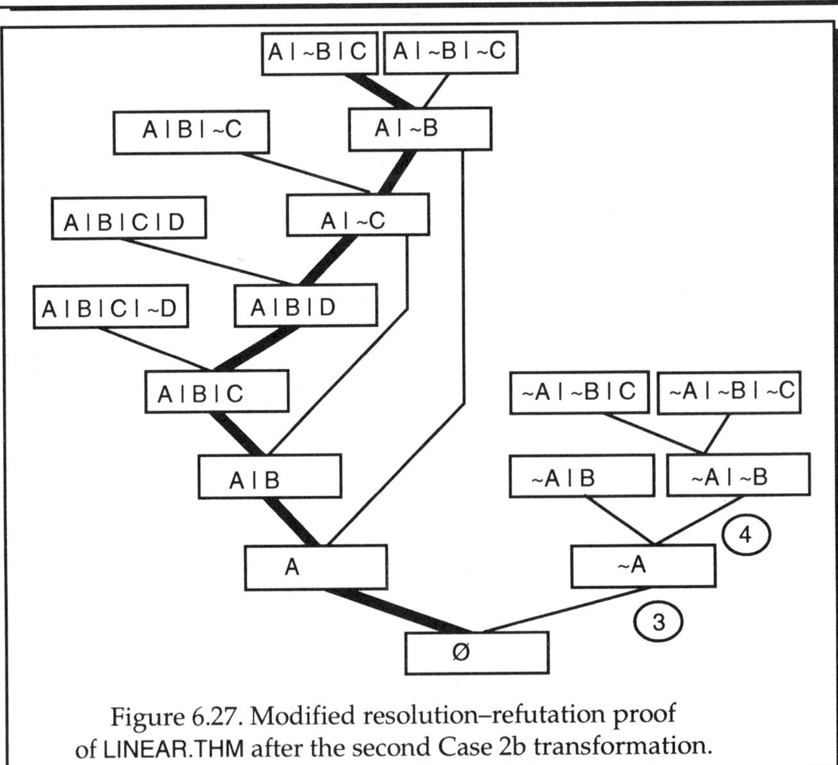


Figure 6.27. Modified resolution–refutation proof of LINEAR.THM after the second Case 2b transformation.

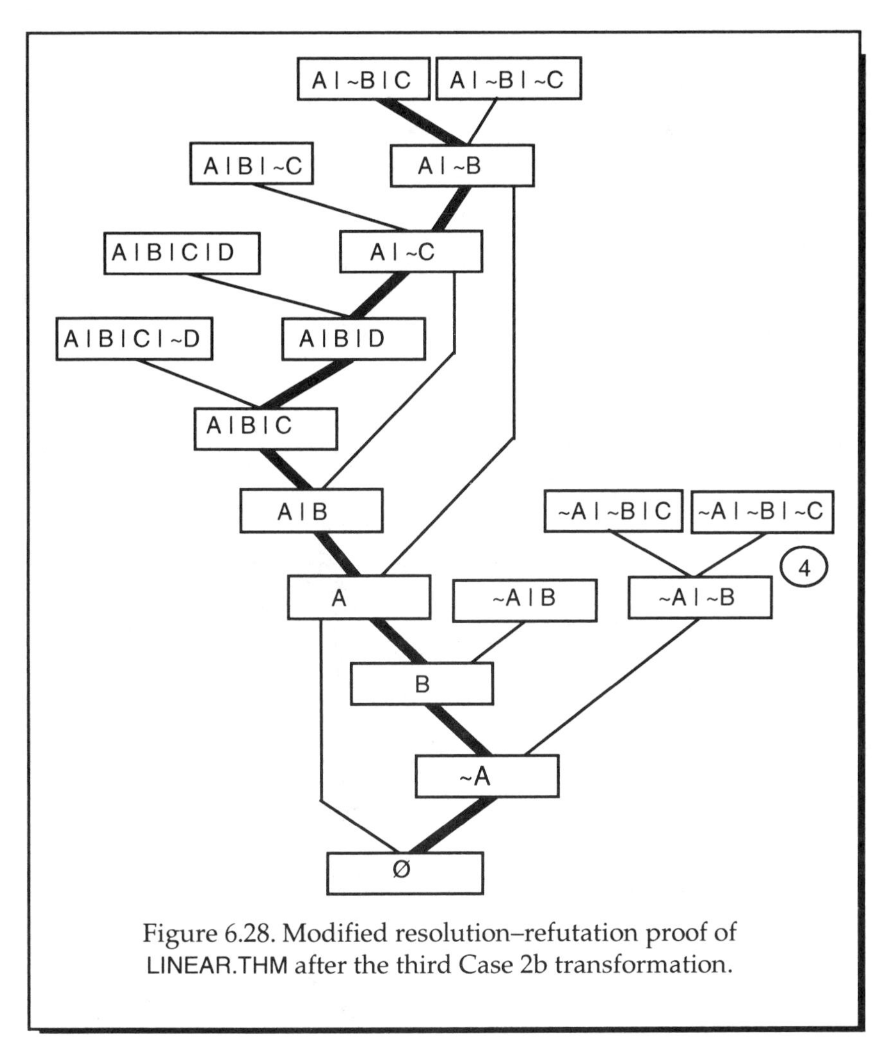


Figure 6.28. Modified resolution–refutation proof of LINEAR.THM after the third Case 2b transformation.

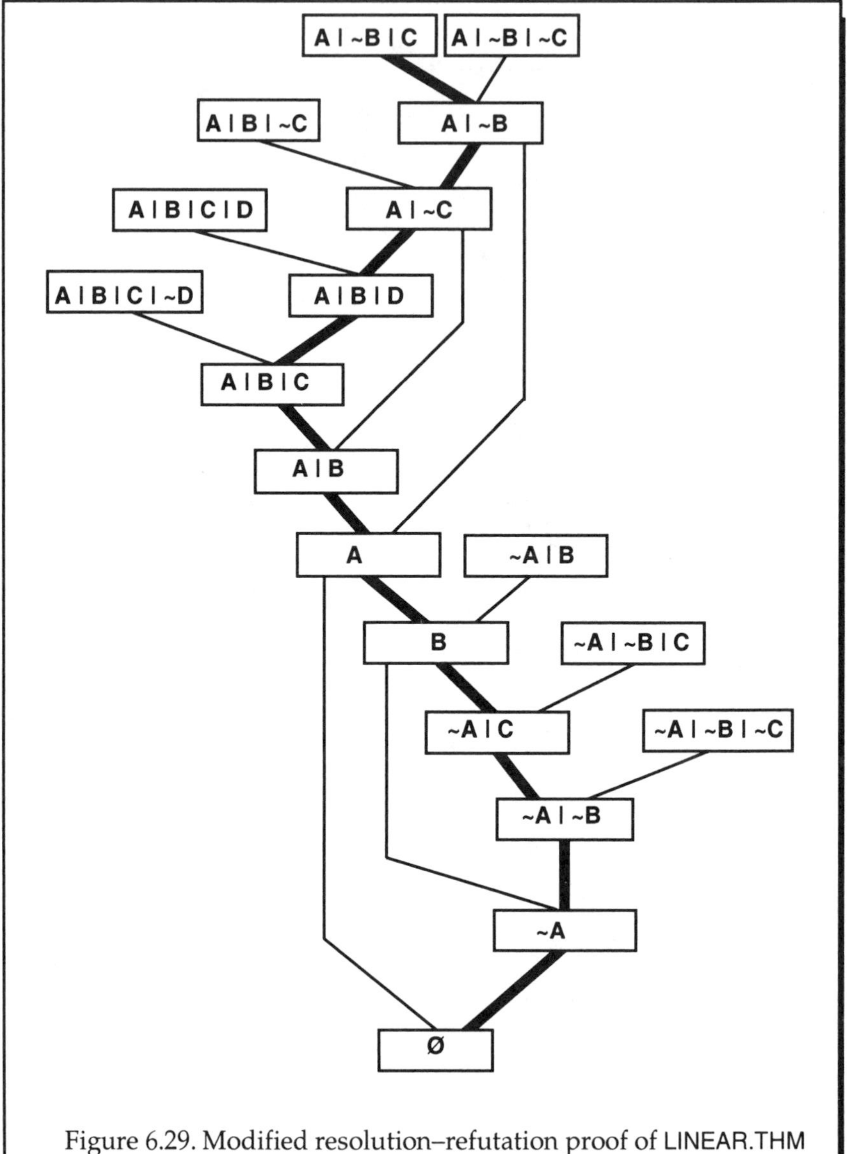


Figure 6.29. Modified resolution–refutation proof of LINEAR.THM after the fourth and final Case 2b transformation.

## 6.5 Restrictions on the Form of Linear Proofs

The transformation of a resolution–refutation proof tree to a linear proof can have surprisingly strong conditions placed on the final structure of the linear proof. First, a lemma is necessary.

**Lemma 6.1. On linear merge-free derivations**
Suppose a linear merge-free derivation D of the clause $C = L \mid L_1$ can be obtained from the clauses $C_1, C_2, ..., C_k$, as shown in Figure 6.30a. Clauses $C_1$ and $C_2$ resolve to yield resolvent $R_1$, clauses $C_3$ and $R_1$ resolve to yield resolvent $R_2$, and so on, where resolvents $R_1, R_2, ..., R_{k-1}$ are not merge clauses (the definition of a linear merge-free derivation). $L_1$ is some disjunction of literals, and L is a single literal. Then, there is a linear merge-free derivation D' of $L_1$ from the same set of clauses $C_1, C_2, ..., C_k$, along with one more clause C = ~L, and, furthermore, clause C is an input clause to the first resolvent in D', as shown in Figure 6.30b.

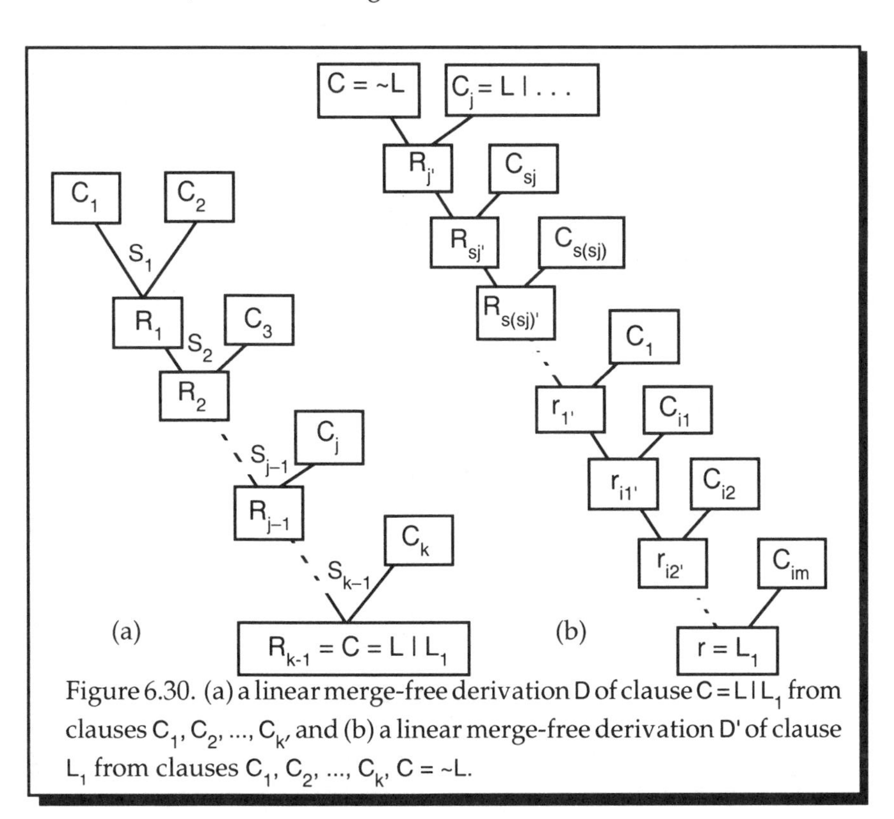


Figure 6.30. (a) a linear merge-free derivation D of clause $C = L \mid L_1$ from clauses $C_1, C_2, ..., C_k$, and (b) a linear merge-free derivation D' of clause $L_1$ from clauses $C_1, C_2, ..., C_k$, C = ~L.

Proof: The proof is by construction. Consider Figure 6.30a. Each $R_i$ in D is the resolvent of $C_{i+1}$ and $R_{i-1}$ (except for i = 1 in which case $R_1$ is the resolvent of $C_1$ and $C_2$) and formed by resolving on some literal in $C_{i+1}$ and its complement in $R_{i-1}$ (again, except for i = 1, in which case the complement comes from $C_1$). With each resolvent $R_i$, let $S_i$ denote the **source clause** of the resolved-on literal in $R_i$ when generating resolvent $R_{i+1}$. Because no resolvent $R_i$ for $1 \le i \le k-1$ is a merge clause, there is a unique source clause for each resolved-upon literal.

Now, construct derivation D' as follows. Note that in derivation D there must be some maximum value of j such that clause $C_j$ has L as one of its literals. Let the top two clauses of D' be C and $C_j$. Then, while j > 1, recursively select the next clause in derivation D' as clause $C_{sj}$ in D, the source clause of the resolved-upon literal that generated resolvent $R_j$ in D, and then set j = sj.

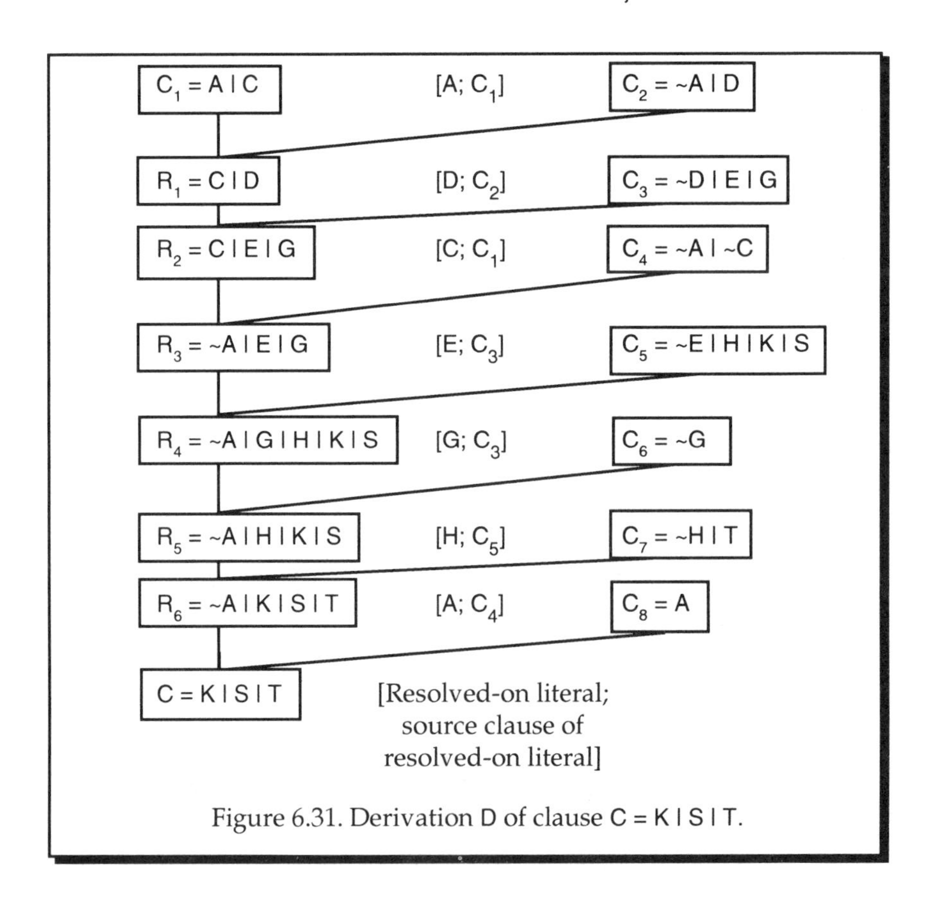


Figure 6.31. Derivation D of clause C = K | S | T.

There may be clauses not selected when this recursive step is completed — that is, when the value of j is 1. Let the unselected clauses in D be in order from the top clause $C_{i1}, C_{i2}, ..., C_{im}$, where m < k. Then, the remaining clauses are resolved in exactly that order, producing resolvents $r_{1'}, r_{2'}, ..., r_{m-k} = L_1$.

**Example 6.8.** Figures 6.31 and 6.32 show an example of this procedure. Using the notation of the lemma, K = L and S | T = $L_1$. The resolvents, the resolved-on literals, and the source clause of each resolved-on literal are shown in Figure 6.31 for clauses $C_1, C_2, ..., C_8$. For example, the resolvent $R_6$ = ~A | K | S | T is formed by resolving clause $C_7$ = ~H | T and resolvent $R_5$ = ~A | H | K | S on the H literal; the H literal of $R_5$ had clause C5 as its source clause.

Continuing our example in Figure 6.32, the top two clauses in the

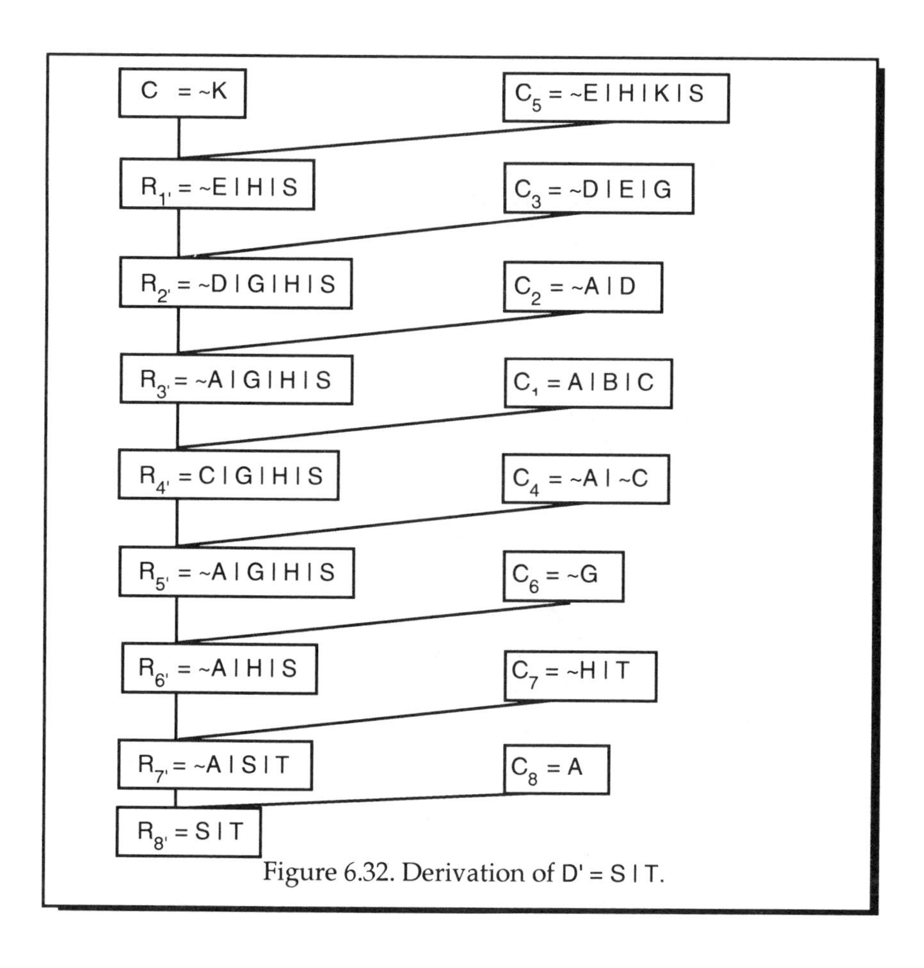

Figure 6.32. Derivation of D' = S | T.

derivation of S | T are C = ~K and $C_5$ = ~E | H | K | S, the resolvent of C and $C_5$ is $R_{1'}$ = ~E | H | S. $R_{1'}$ is next resolved with clause $C_3$ = ~D | E | F | G. The resolved-on literal is E and the resolvent is $R_{2'}$. $R_{2'}$ and $C_2$ are resolved together next to produce $R_{3'}$, which is then resolved with $C_1$ to produce $R_{4'}$.

In this example, clauses $C_4$, $C_6$, $C_7$, and $C_8$ were not selected by the recursive procedure described in the lemma, and they are now resolved in this order, producing the resolvents $R_{5'}$, $R_{6'}$, $R_{7'}$, and $R_{8'}$ = S | T and the tree, as shown in Figure 6.32.

We now present a corollary to Lemma 6.1. The corollary is a minor variation of the lemma.

**Corollary 6.1. (To Lemma 6.1.) on linear merge-free derivations**
Suppose a linear merge-free derivation D of the clause C = L | $L_1$ can be obtained from the clauses $C_1$, $C_2$, ..., $C_k$, as shown in Figure 6.30a. Clauses $C_1$ and $C_2$ resolve to yield resolvent $R_1$, clauses $C_3$ and $R_1$ resolve to yield resolvent $R_2$, and so on, where resolvents $R_1$, $R_2$, ..., $R_{k-1}$ are not merge clauses. $L_1$ is some disjunction of literals, and L is a single literal. Then, there is a linear merge-free derivation D' of $L_1$ | $L_2$ from the set of clauses $C_1$, $C_2$, ..., $C_k$, C = ~L | $L_2$, where $L_2$ is a single literal, and, furthermore, clause C is an input clause to the first resolvent.

A **linear-merge proof** is a linear proof in which if resolution $R_j$ has resolution $R_i$ for a parent, and if j > i+1, then $R_i$ is a merge clause.

**Theorem 6.3. Every theorem has a linear-merge proof**
Proof: The proof is by construction. Use Algorithm 6.2 with a modification to Step 3 as described following this paragraph when (see Figure 6.21) the tree rooted at resolvent $R_k$ conforms to Case 2b and resolvent $R_{k-1}$ is not a merge clause. If $R_{k-1}$ is a merge clause, the transformation described in Figure 6.23 can be made. If $R_{k-1}$ is not a merge clause, the transformation shown in Figure 6.23 cannot be made because that would mean resolvent $R_{k+2}$ in Figure 6.23 would have an input clause, $R_{k-1}$, which is not a merge clause or base clause and which comes from a resolvent two or more levels higher in the linear tree. A different, more complex transformation is necessary that makes use of Corollary 6.1.

Now, when $R_{k-1}$ is not a merge clause, as it is in the case of concern, either the derivation of $R_{k-1}$ from the base clauses is merge-free, or there is some

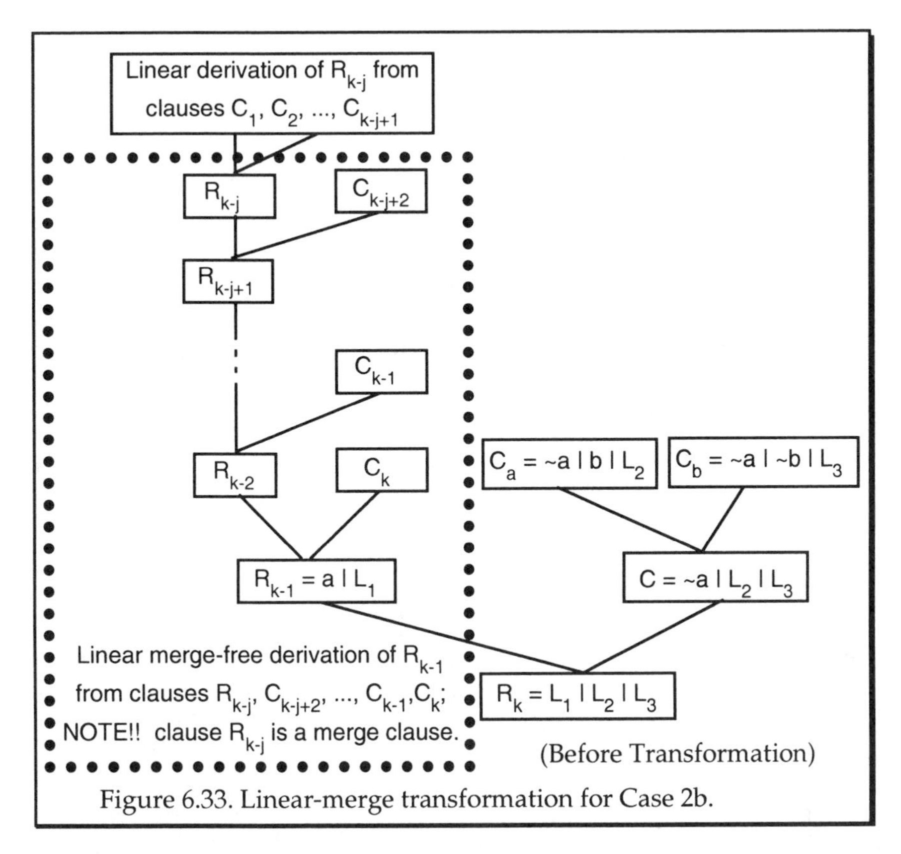

Figure 6.33. Linear-merge transformation for Case 2b.

clause $R_{k-j}$ for $j > 1$ that is a merge clause and $R_{k-j+1}, R_{k-j+2}, ..., R_{k-1}$ are not merge clauses. In the first case, the derivation of $R_{k-1}$ from $C_1, C_2, ..., C_k$ is merge-free. In the second case, the derivation of $R_{k-1}$ from the clauses $R_{k-j}, C_{k-j+2}, ..., C_{k-1}, C_k$ is merge-free.

Rather than the transformation shown in Figure 6.23, the transform shown in Figure 6.33 should be performed. The transformation takes advantage of Corollary 6.1 to eliminate the literal ~a in the clause $R_{k+1} = \sim a \mid L_1 \mid L_2 \mid L_3$, yielding the clause $R_{k+2} = L_1 \mid L_2 \mid L_3$.

**Example 6.9.** Figures 6.34–6.36 present an example of this theorem. Theorem LINEAR_MERGE.THM is presented in Figure 6.34. A resolution-refutation proof is shown in Figure 6.35. The derivation of resolvent $R_4 = \sim A$ in Figure 6.35 is linear, but a transformation of the type illustrated in Figure 6.33 must be made, where the $R_k$, $R_{k-1}$, C, $C_A$, and $C_B$ of Figure 6.33 correspond to $R_5 = \emptyset$, $R_4 = \sim A$, $C_9 = A$, $C_6 = A \mid F$ and $C_7 = A \mid \sim F$, respectively,

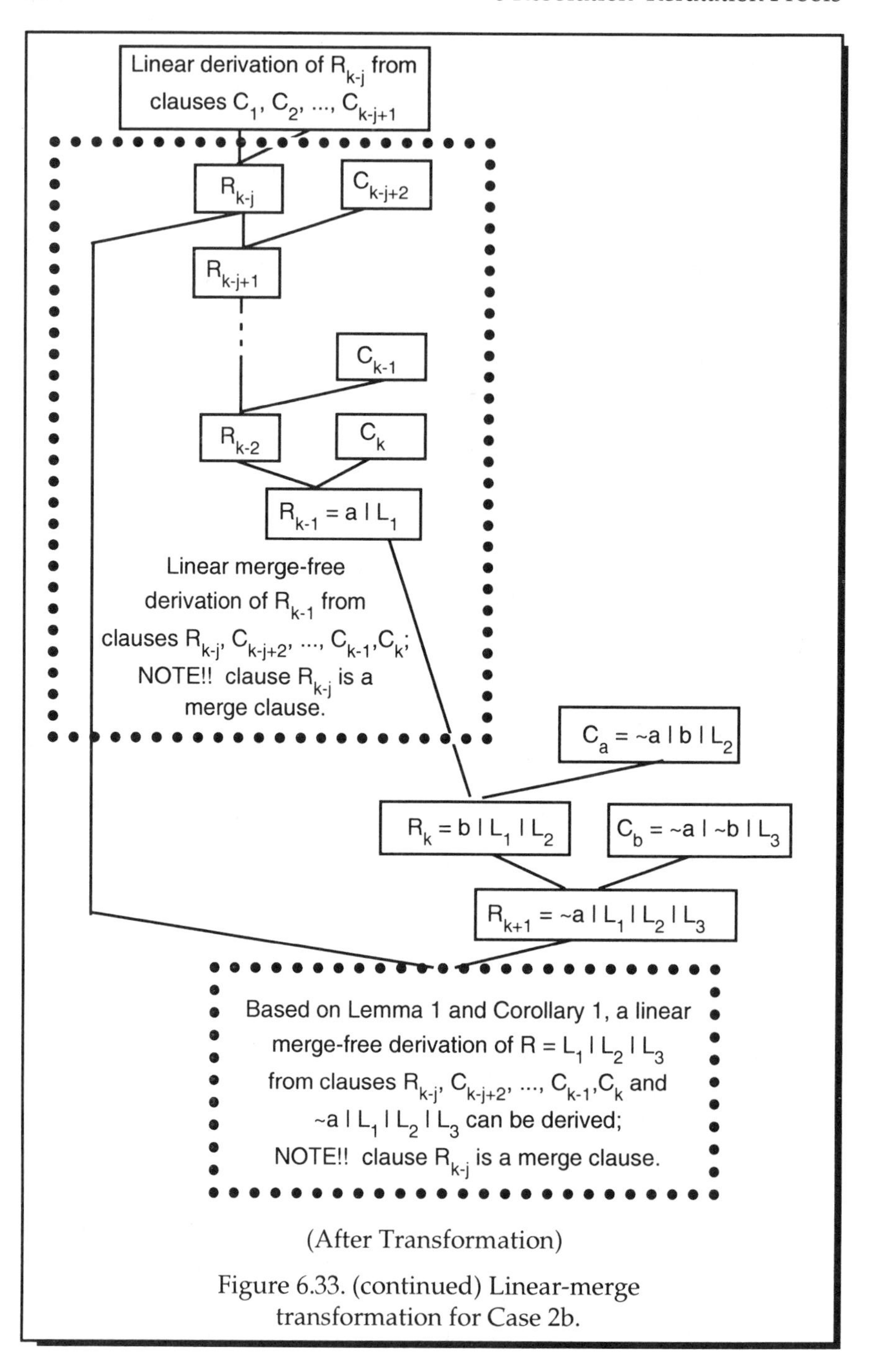


(After Transformation)

Figure 6.33. (continued) Linear-merge transformation for Case 2b.

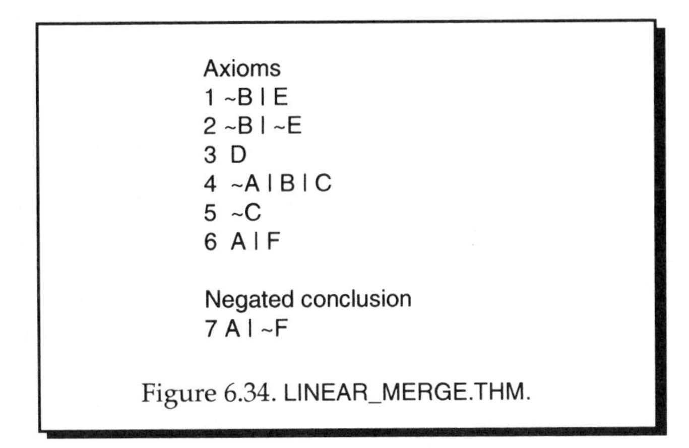


Figure 6.34. LINEAR_MERGE.THM.

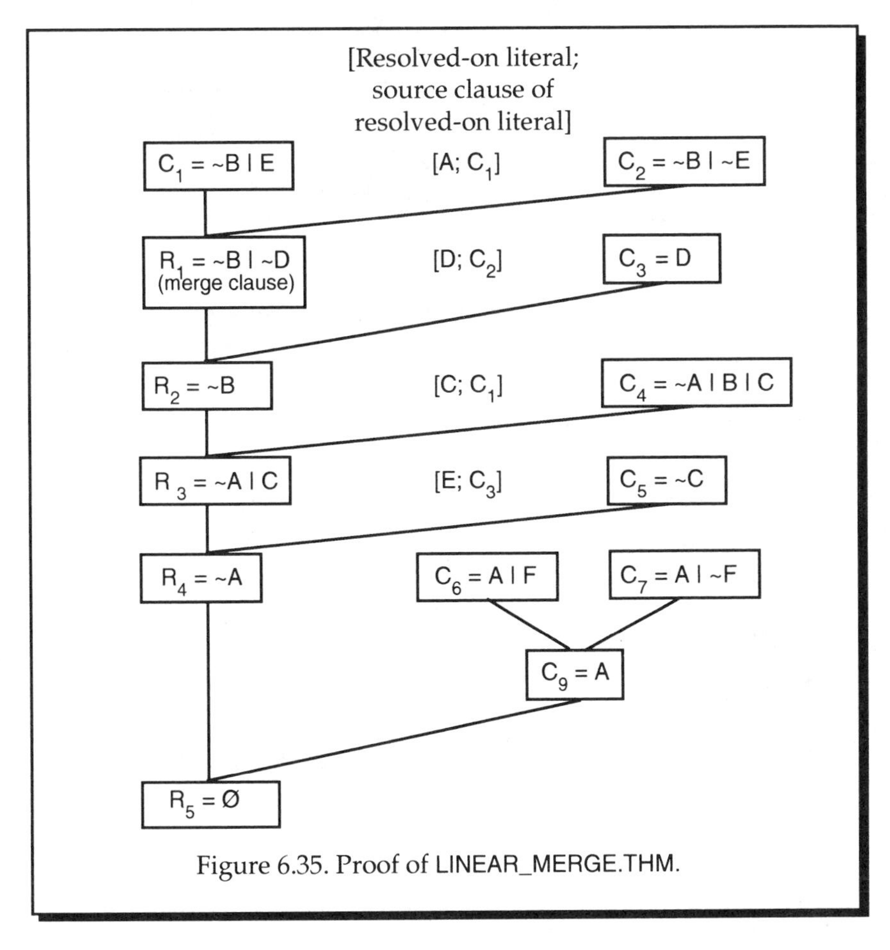


Figure 6.35. Proof of LINEAR_MERGE.THM.

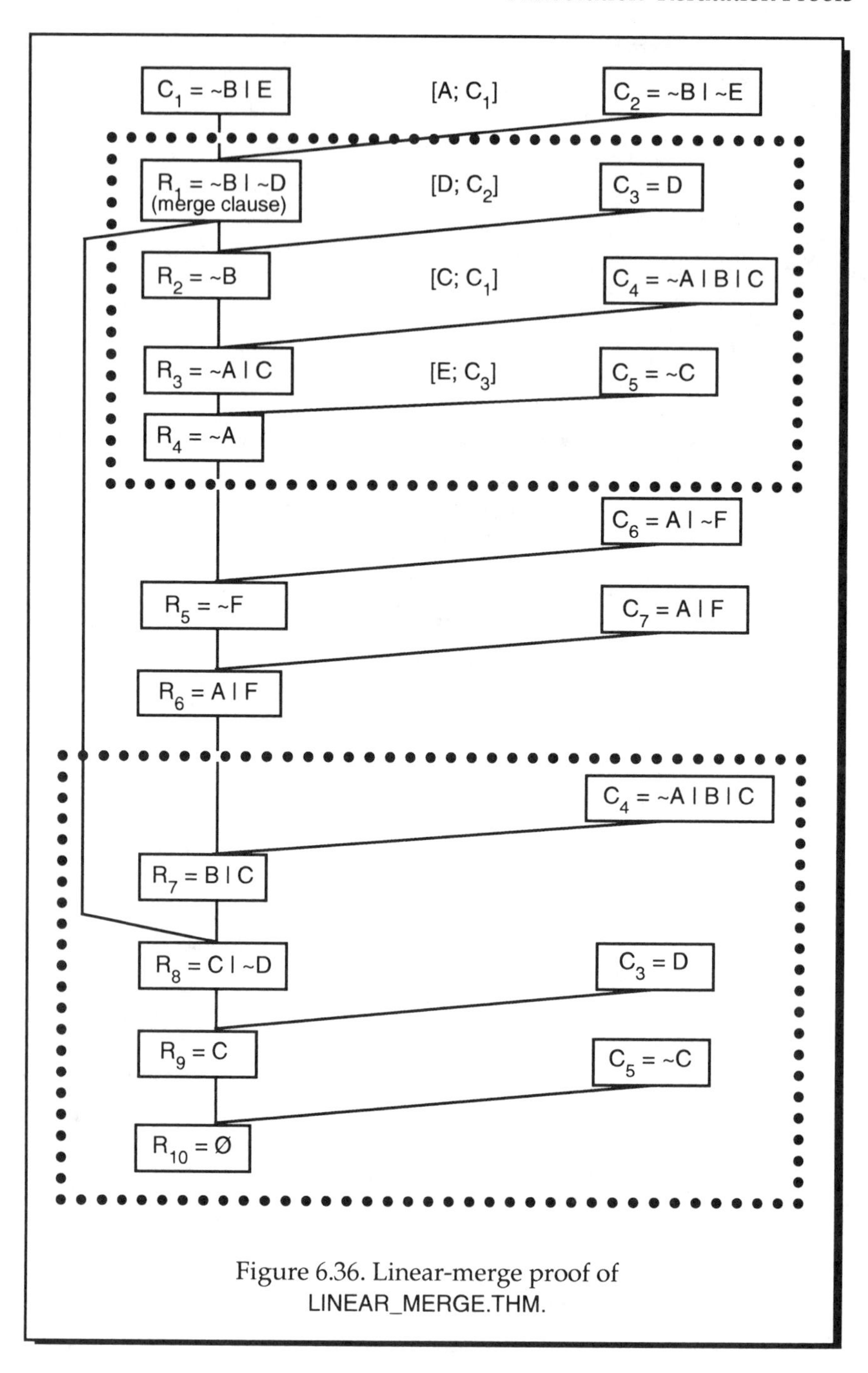


Figure 6.36. Linear-merge proof of LINEAR_MERGE.THM.

in Figure 6.35. Note that $R_4$, $R_3$, and $R_2$ are not merge clauses but that $R_1$ is, and thus resolvents $R_{k-j}$, $R_{k-j+1}$, ..., $R_{k-2}$, $R_{k-1}$ in Figure 6.33 correspond to $R_2$, $R_3$, and $R_4$ in Figure 6.35.

**Lemma 6.2. A clause from the negated conclusion of a theorem must appear in any resolution–refutation proof of the theorem**
Proof: If such a clause does not appear, then the clauses that constitute the axioms contain a contradiction.

A **linear-nc proof** is a linear proof having a clause from the negated conclusion as one of the input clauses to the first resolution.

**Theorem 6.4. Every theorem has a linear-nc proof**
Proof: This is left for the reader to find (see Exercise 6.9).

**Theorem 6.5. Every theorem has a linear-merge-nc proof**
Proof: This is left for the reader to find (see Exercise 6.9).

## 6.6 The Lifting Lemma

The **lifting lemma**, which will only be stated, allows the results presented in Sections 6.4 and 6.5 to be extended to the case of first-order predicate calculus

**Lemma 6.3. The lifting lemma**
Given a set of clauses S and another set of clauses S' that are ground instances of clauses in S, and given that $I_1'$, $I_2'$, ..., $I_n'$ is a derivation of some clause C' from the set S', then there exists a derivation $I_1$, $I_2$, ..., $I_n$ = C from the set S and a sequence of substitutions, $ø_1$, $ø_2$, ..., $ø_n$, such that $I_1ø_1 = I_1'$, $I_2ø_2 = I_2'$, ..., $I_nø_n = I_n' = C'$.

This lemma justifies extending the transformations presented in Figures 6.22 and 6.23 to the more general case of first-order predicate calculus. It then follows that Algorithm 6.2 is valid for theorems expressed in first-order predicate calculus. It is left to the reader to see how factoring should be handled.

**Example 6.10.** Let S be the set of clauses for Q07B0.THM presented in Figure 6.1, and let S' be ground instances of clauses in S as shown in Figure 6.37a. Let the derivation of the NULL clause in S' be as shown in Figure 6.37b. Then,

```
    4'  B(b,c,Ext(b,c,d,e))
   20'  ~Equal(a,Ext(b,c,d,e)) | Equal(Ext(b,c,d,e),a)
   24'  ~Equal(Ext(b,c,d,e),a) | ~B(b,c,Ext(b,c,d,e)) | B(b,c,a)
   39'  Equal(a,Ext(b,c,d,e))
   40'  ~B(b,c,a)

   41': (40'a,24'c)  ~Equal(Ext(b,c,d,e),a) | ~B(b,c,Ext(b,c,d,e))
   42': (41'a,20'b)  ~Equal(a,Ext(b,c,d,e)) | ~B(b,c,Ext(b,c,d,e))
   43': (42'b,4'a)  ~Equal(a,Ext(b,c,d,e))
   44': (43'a,39'a) Ø
```

Figure 6.37. (a) Ground instances of clauses in S, and (b) a resolution–refutation proof of Q07B0.THM.

comparing clauses 41, 42, 43, and 44 in Figure 6.2 and clauses 41', 42', 43', and 44' in Figure 6.37b, the sequence of substitutions is: $ø_1$ = {Ext(b,c,d,e)/x}, $ø_2$ = {Ext(b,c,d,e)/x}, $ø_3$ = {ø/ø}, $ø_4$ = {ø/ø}.

## 6.7 Linear Proofs and Factoring

The material presented so far on linear proofs has not considered factoring. In Figure 6.20, two cases of derivations rooted at some clause $R_k$ are considered. The third case, where clause C is a binary factor, as shown in Figure 6.38, was not considered. If C is a binary factor, there are two subcases, shown in Figure 6.39 as Case 3a and Case 3b. In the first subcase,

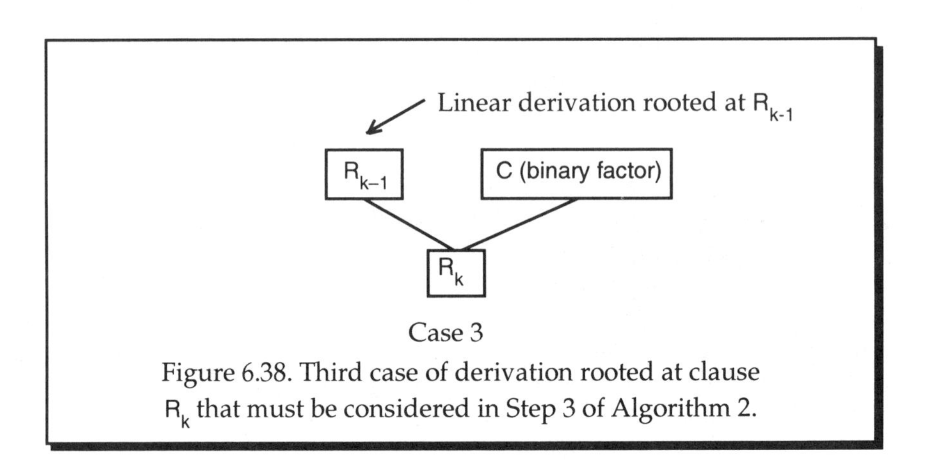


Figure 6.38. Third case of derivation rooted at clause $R_k$ that must be considered in Step 3 of Algorithm 2.

clause C' contains two literals that were factored producing a literal that was not resolved with a literal in $R_{k-1}$. In the second subcase, clause C" contains two literals that were factored producing a literal that was resolved with a literal in $R_{k-1}$.

In the first subcase, the transformation in Figure 6.40 is applicable. In the second subcase, the transformation shown in Figure 6.41 is applicable. Here, feed-forward is necessary from $R_{k-1}$ to $R_{k+1}$.

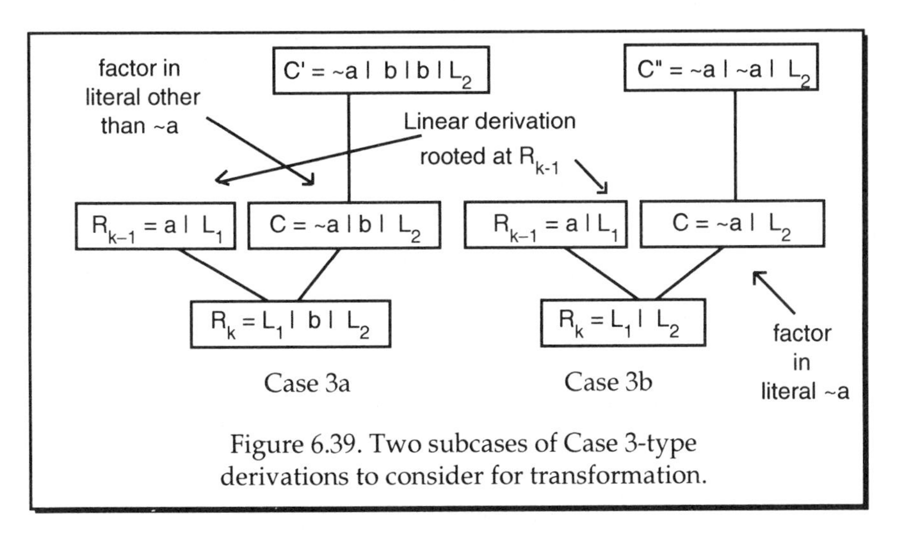


Figure 6.39. Two subcases of Case 3-type derivations to consider for transformation.

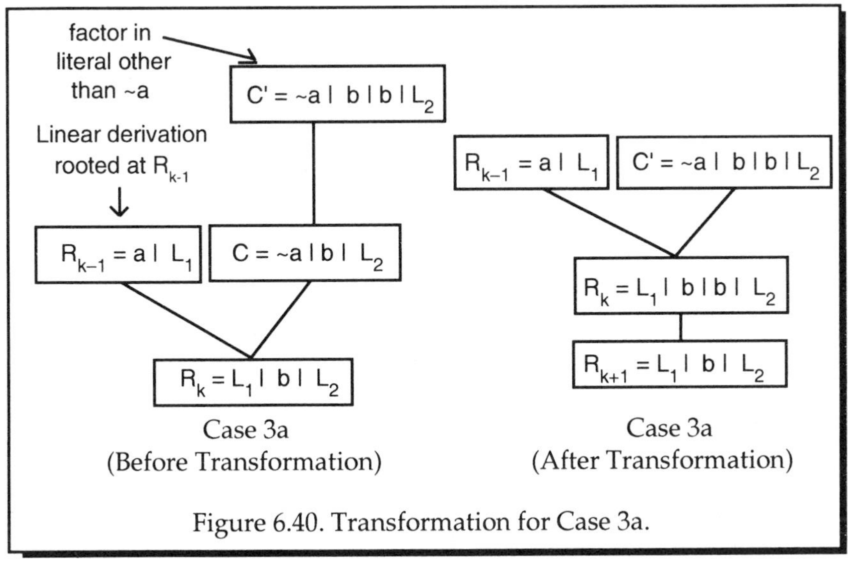


Figure 6.40. Transformation for Case 3a.

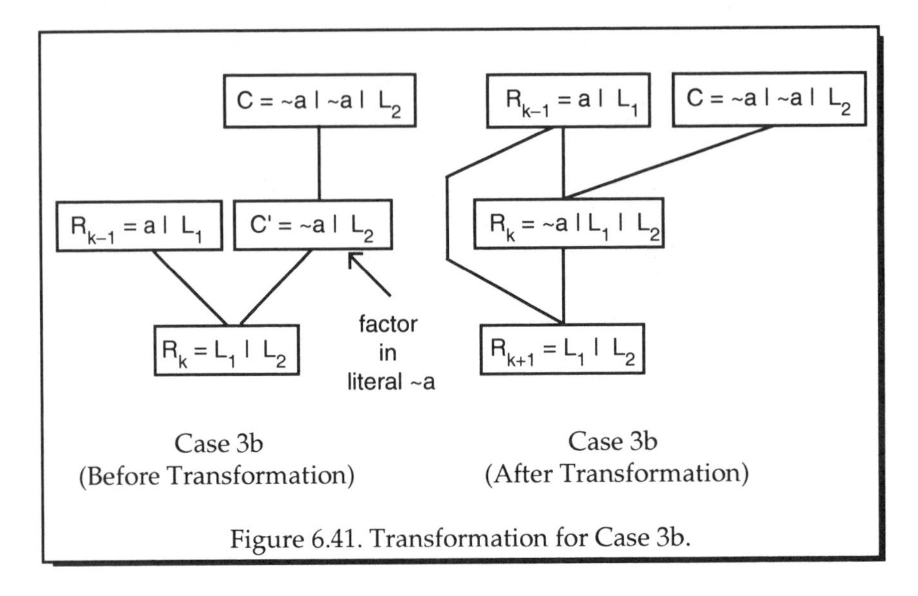


Figure 6.41. Transformation for Case 3b.

## Exercises for Chapter 6

6.1. What was the original wff that, when compiled into clauses, yielded clauses 8, 9, and 10 of STARK017.THM shown in Figure 6.7?

6.2. Beginning with the semantic tree (see Exercise 5.4) for EX2.THM in Figure 5.9, use Algorithm 6.1 to obtain a resolution–refutation proof.

6.3. Using the semantic tree constructed for FACT.THM in Exercise 5.5 and Algorithm 6.1, construct a resolution–refutation proof.

6.4. Transform the resolution–refutation proof tree in Figure 6.12 to linear form.

6.5. Give an algorithm that finds a set of atoms that can be used to generate a closed semantic tree when given a resolution–refutation proof. Use your algorithm to find atoms that can generate a closed semantic tree for Q07B0.THM.

6.6. Suppose Algorithm 6.2 is applied to a theorem which has a resolution–refutation proof of depth D. What is the maximum depth of the resulting linear form?

6.7. A **vine-form proof** is a linear proof with the added restriction that each resolution in the proof has at least one base clause as an input. Provide a simple example that shows that not all theorems have a vine-form proof.

6.8. Is the following statement true or false? If a theorem has a proof without merge clauses, then it also has a vine-form proof. Discuss your answer.

6.9. Prove Theorems 6.4 and 6.5.

6.10. Using the lifting lemma, show that the transformations shown in Figures 6.22 and 6.23 are also valid for first-order predicate calculus.

6.11. Using the transform shown in Figure 6.33, transform the resolution–refutation proof shown in Figure 6.43 of EX4.THM (Figure 6.42) to a linear-merge proof. Use the path shown in bold for the principal path. Show how Corollary 6.1 applies.

6.12. The theorem SQROOT.THM in Figure 6.44 says that the square root of a prime number is irrational. Its resolution–refutation proof is given in Figure 6.45. Transform the proof to linear form.

Axioms:
1 P I T    2 P I ~T
3 ~P I Q    4 ~Q I R
5 ~R I S

Negated conclusion
6 ~R I ~S

Figure 6.42. EX4.THM.

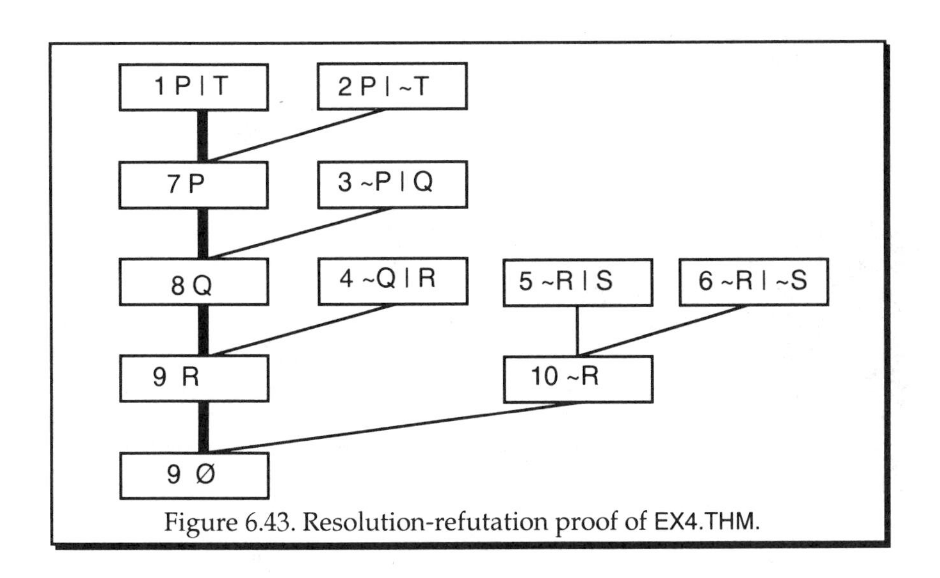

Figure 6.43. Resolution-refutation proof of EX4.THM.

Theorem - The square root of a prime is irrational. P(x): x is a prime, T(x,y,z): x times y equals z, D(x,y): x divides y, sq(x) is the square function, and SK(x,y) is a Skolem function. Constants: a, b, 1, p.

```
Axioms
1  ~P(x) | ~T(y,z,u) | ~D(x,u) | D(x,y) | D(x,z)
2  ~T(x,y,z) | ~T(z,u,v) | ~T(x,w,v) | T(u,y,w)
3  ~D(x,y) | ~D(y,z) | D(x,z)
4  ~D(x,a) | ~D(x,b) | Equal(x,1)
5  ~D(x,y) | T(x,SK(x,y),y)
6  ~T(x,y,z) | D(x,z)
7  T(x,x,sq(x))
8  P(p)                  ; if p is prime and not 1, then for all x and y
9  ~Equal(p,1)           ; p times the square of x is not equal to the

Negated conclusion       ; square of y, i.e., the square root of p is
10  T(p,sq(b),sq(a))     ; irrational
```

Figure 6.44. SQROOT.THM, a theorem on prime numbers.

```
11: (9a,4c)   ~D(p,a) | ~D(p,b)
12: (8a,1a)   ~T(x,y,z) | ~D(p,z) | D(p,x) | D(p,y)
13: (12a,7a)  ~D(p,sq(x))  | D(p,x)
14: (13a,6b)  ~T(p,x,sq(y)) | D(p,y)
15: (14a,10a) D(p,a)
16: (13b,11b) ~D(p,sq(b)) | ~D(p,a)
17: (16b,15a) ~D(p,sq(b))
18: (10a,2c)  ~T(p,x,y) | ~T(y,z,sq(a)) | T(z,x,sq(b))
19: (18c,6a)  ~T(p,x,y) | ~T(y,z,sq(a)) | D(z,sq(b))
20: (19a,5b)  ~T(x,y,sq(a)) | D(y,sq(b)) | ~D(p,x)
21: (20b,3b)  ~T(x,y,sq(a)) | ~D(p,x) | ~D(z,y) | D(z,sq(b))
22: (21bc)    ~T(x,x,sq(a)) | ~D(p,x) | D(p,sq(b))
23: (22c,17a) ~T(x,x,sq(a)) | ~D(p,x)
24: (23a,7a)  ~D(p,a)
25: (24a,15a) Ø
```

Figure 6.45. Proof of SQROOT.THM.

# 7 HERBY: A Semantic–Tree Theorem Prover

It was shown in Chapter 5 that, in theory, a closed canonical semantic tree can be constructed for any unsatisfiable set of clauses. In practice, constructing closed canonical semantic trees is not a very effective procedure because often far too many atoms must be selected before a closed tree is obtained. However, as was also discussed in Chapter 5, semantic trees need not be canonical and, when this is the case, a stronger theorem prover can be designed. HERBY is just such a prover, although it is still considerably weaker than programs that use resolution–refutation.

In this chapter, the heuristics used by HERBY are described. The first group, described in Section 7.1, is used to select atoms in a noncanonical order. The second group, described in Section 7.2, improves HERBY's effectiveness but is not directly concerned with atom selection.

## 7.1 Heuristics for Selecting Atoms

HERBY is an implementation of Algorithm 5.1. It uses five atom selection heuristics and one constant selection heuristic in the order presented to help select atoms effectively. A sixth atom selection heuristic permits the user to select atoms. This section describes these heuristics.

In the following, let N denote a node at which an atom is to be selected. HERBY stores clauses assigned to the nodes on the path from the root to and including node N on a list called **clist**. The base clauses are at the head of this list. There is space for ten thousand clauses. HERBY uses a second list, **unit_list**, to store up to fifteen hundred unit clauses generated by ASH4.

**Atom selection heuristic 1 (ASH1). Generate atom from 1–1 resolvable pair** Search clist for two clauses that are a **1–1 resolvable pair**. Two clauses,

C1 = L1 and C2 = ~L2, are a 1-1 resolvable pair if both are unit clauses and they resolve to yield the NULL clause. If such a pair is found, substitute the mgu of the two literals, L1 and L2, into the positive literal, L1, and use the resulting literal, if it is a ground instance, for the next atom. If it is not a ground instance, ground it by substituting a constant for all variables according to the constant selection heuristic (see page 91). The two nodes that are successors of N will both fail.

**Example 7.1.** Suppose clist contains the unit clauses P(f(a),x) and ~P(y,a). They form a 1–1 resolvable pair yielding P(f(a),a). This is a ground instance and becomes the next atom. As a second example, suppose T(N) contains P(f(x),y) and ~P(u,f(v)). Their mgu is {f(x)/u, f(v)/y}, giving P(f(x),f(v)), and when a constant, say "a," is substituted for the x and v in this clause, their most common instance P(f(a),f(a)) becomes the next atom selected.

**Atom selection heuristic 2 (ASH2). Generate atoms from 1–1–2 hyper-resolvable trio**

Search clist for three clauses C1 = L1, C2 = L2, and C3 = L3 | L4 that are a **1–1–2 hyperresolvable trio**. Three clauses C1, C2, and C3 are a 1–1–2 hyper-resolvable trio if C1 resolves on literal L1 with one of the literals, say L3, of C3 to yield some unit clause C4 that in turn resolves with clause C2 on literal L2 to yield the NULL clause. If such a trio is found, form the resolvent of C1

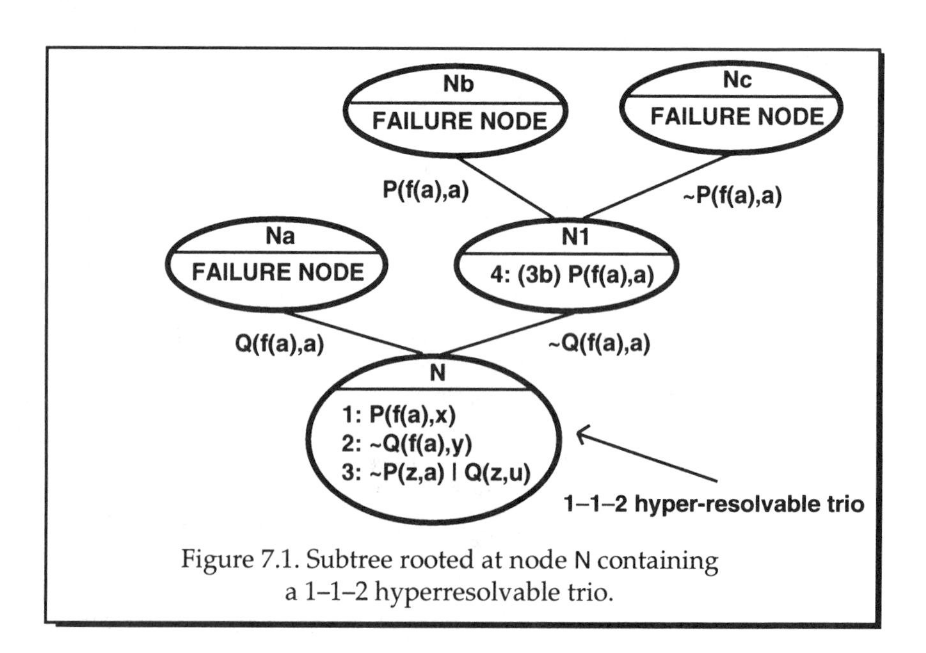


Figure 7.1. Subtree rooted at node N containing a 1–1–2 hyperresolvable trio.

and C3 by resolving on literal L3 of C3 and L1 of C1 forming unit clause C4. Then, using the mgu of C2 and C4, determine the most common instance of C2 and C4 and select it, if it is a ground instance, for the next atom. If it is not a ground atom, ground it by substituting the same constant for all variables as dictated by the constant selection heuristic. The other atom selected is the most common instance of literals L1 and L3.

One child of N will be a failure node, failing because of C2. At the other child, its children will fail because of C1 and C3.

**Example 7.2.** Suppose, as shown in Figure 7.1, that clist contains: 1: P(f(a),x), 2: ~Q(f(a),y), 3: ~P(z,a) | Q(z,u). These clauses form a 1–1–2 hyperresolvable trio. Clauses 1 and 3 resolve with the mgu {f(a)/z,a/x} to yield 4: Q(f(a),u), which in turn resolves with clause 2 with the mgu {y/u} to yield the NULL clause. The atoms selected are then Q(f(a),a) and P(f(a),a). One child of N, node Na, will be a failure node. The children, node Nb and node Nc, of the other child, N1, will both fail, too.

**Atom selection heuristic 3 (ASH3). Generate atoms from 1–1–1–3 hyperresolvable quad**

Search clist for four clauses C1 = L1, C2 = L2, C3 = L3, and C4 = L4 | L5 | L6 that are a **1–1–1–3 hyperresolvable quad**. Four clauses C1, C2, C3, and C4 are a 1–1–1–3 hyperresolvable quad if C1 resolves with the first literal of C4 to yield some clause C5 whose first literal resolves with C2 to yield some clause C6 that in turn resolves with C3 to yield the NULL clause. If such a quad is found, form the resolvent of C1 and C4 by resolving on the first literal of C4 to form clause C5. Then, form the resolvent of C2 and C5 by resolving on the first literal of C5 to form clause C6. Next, form the resolvent of C3 and C6 to yield the NULL clause. The most common instance of clauses C3 and C6 is selected as the next atom. Select as a second atom the most common instance of L1 and the literal of C4 that resolves with L1. Select as the third atom the most common instance of L2 and L5. Ground ungrounded atoms as done with ASH1 and ASH2 using the constant selection heuristic.

One child of N will be a failure node, failing because of C3. At the other child, there will be two children, one of which will fail because of C2. At the other, there will be two children, and they will fail because of C1 and C4.

**Example 7.3.** Suppose, as shown in Figure 7.2, that clist contains: 1: P(f(a),x), 2: ~Q(f(a),y), 3: ~R(z,u), 4: ~P(v,a) | Q(v,w) | R(a,f(a)). These clauses form a 1–1–1–3 hyperresolvable quad. Clauses 1 and 4 resolve together with the mgu {f(a)/v,a/x} to yield clause 5: Q(f(a),w) | R(a,f(a)), which in turn resolves with

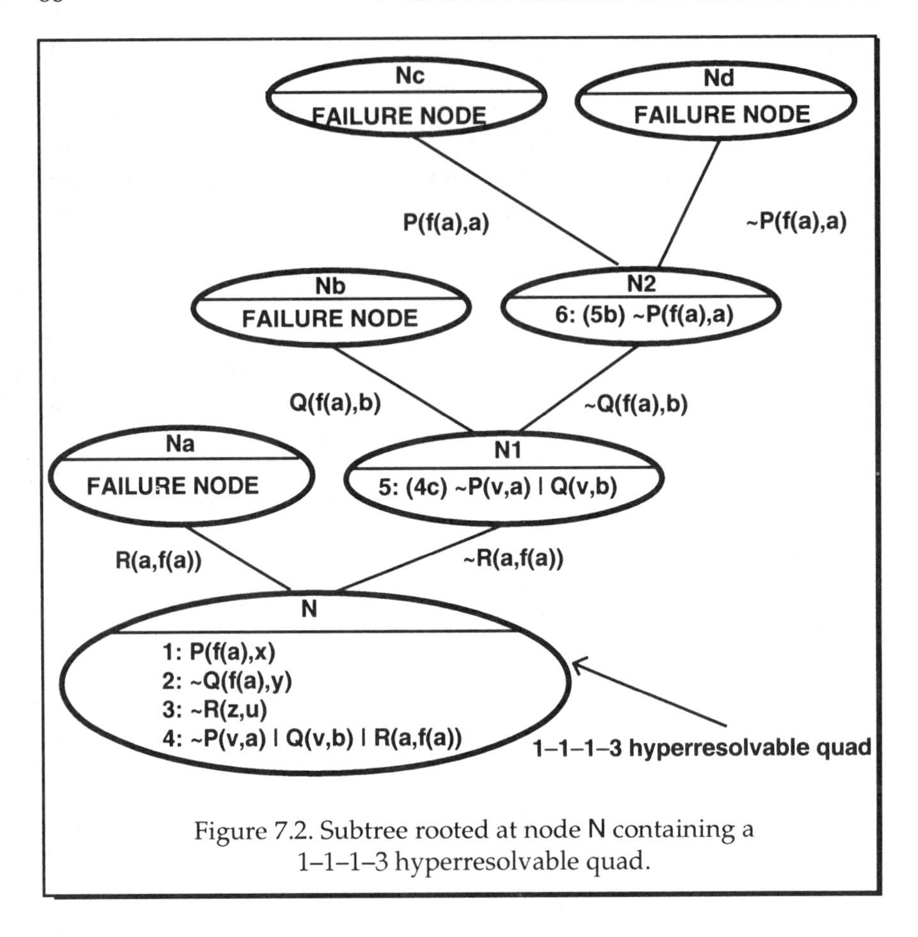


Figure 7.2. Subtree rooted at node N containing a 1–1–1–3 hyperresolvable quad.

clause 2 with the mgu {y/w} to yield clause 6: R(a,f(a)), which in turn resolves with clause 3 with the mgu {a/z,f(a)/u} to yield the NULL clause. Thus, ASH3 selects for atoms R(a,f(a)), ~Q(f(a),b), and P(f(a),a). One child of N, Na, is a failure node. At the other child of N — that is, node N1 — there are two children. One fails clause 2. At the other child — that is, node N2 — there are two children that fail clause 1 and clause 4.

Two clauses C1 = L1 and C2 = L2 | L3 are a **unit-generating pair** if C1 is a unit clause and C2 has two literals, and C1 and C2 resolve to yield a **1–2 unit clause**. Three clauses C1 = L1, C2 = L2, and C3 = L3 | L4 | L4 are a **unit-generating trio** if C3 resolves with C1 to form a clause that resolves with the single literal of C2 to yield a **1–1–3 unit clause**.

**Atom selection heuristic 4 (ASH4). Generate atoms using ulist clauses**
ASH4 generates all 1–2 unit clauses and all 1–1–3 unit clauses from clauses on clist. As each new unit clause U is formed, it is resolved with each previously generated unit clause that has been saved on unit_list to see if the pair yields the NULL clause. If so, the next atom selected is a most common instance of the two unit clauses, grounded if necessary by the constant selection heuristic. In addition to this atom, atoms that are the most common instances of the unit-generating clauses leading to U are also selected as atoms, again, grounded if necessary by the constant selection heuristic

This heuristic does not guarantee a closed semantic tree at node N because the unit clauses U and V that resolve may be on different paths in the tree. However, if the clauses that formed the unit clauses are all on clist, the semantic tree will be closed at N, as the following example shows. A high percentage of the time this is the case. ASH4 has been found to be a powerful heuristic and is often the key to proving the more difficult theorems.

**Example 7.4.** Suppose T(N) contains: 1: P(f(a),x), 2: ~R(y,z), 3: ~P(u,a) | Q(u,a) | R(a,f(a)), 4: P(a,f(a)), and 5: ~P(v,f(a)) | ~Q(f(a),v). Clauses 1, 2, and 3 resolve to yield a 1–1–3 unit clause. Specifically, clauses 1 and 3 resolve together with the mgu m1 = {a/x,f(a)/u} to yield clause 6: Q(f(a),a) | R(a,f(a)), which in turn resolves with clause 2 with the mgu m2 = {a/y,f(a)/z} to yield a 1–1–3 unit clause 7: Q(f(a),a). Clauses 4 and 5 resolve together with the mgu m3 = {a/v} to yield a 1–2 unit clause 8: ~Q(f(a),a). Clauses 7 and 8 resolve together with mgu m4 = {Ø/Ø} to yield the NULL clause. Thus, Q(f(a),a) is selected as the next atom and then P(f(a),x) and R(a,f(a)). There is a 1–1 resolvable pair at one child and a 1–1–2 hyperresolvable trio at the other.

**Atom selection heuristic 5 (ASH5). Generate atoms using clist and ulist clauses**
If no atom is found by the first four heuristics, HERBY panics and tries to find any old atom! ASH5 has three subheuristics, each of which attempts to find an atom. An attempt is made to alternate the selection of negated and unnegated literals. The heuristic also mixes up the order of selecting ASH5a, ASH5b, and ASH5c somewhat randomly. There is also some overlap in the clauses considered by the three subheuristics — lots of possibilities for improvements here!

ASH5a. Scan clist for a literal that yields an atom.

There are four subheuristics.

i. Scan clist, excluding base clauses, for a unit ground clause opposite in logical value from the last atom selected. Select it as the next atom.
ii. Scan clist, excluding base clauses, for a grounded unit clause of the

same logical value as the last atom selected. Select it as the next atom.

iii. Scan clist, excluding base clauses, for a unit clause with one variable. Ground it and select it as the next atom.

iv. For all clauses on clist with two literals, first find a literal with no variables and then with one variable. Ground the literal if necessary and select it as the next atom.

ASH5b. Scans unit_list for a unit clause that yields an atom.
There are four subheuristics.

i. Scan unit_list for a unit clause not on clist, and opposite in logical value from the last atom selected. Ground it if necessary and select it as the next atom.

ii. Scan unit_list for a unit clause not on clist, and of the same logical value as the last atom selected. Ground the unit clause if necessary and select it as the next atom.

iii. Scan unit_list for a unit clause with one variable. Assign a constant to the the variable, and select the clause as the next atom.

iv. Scan unit_list for a unit clause with more than one variable. Ground the unit clause and select it as the next atom.

ASH5c. Scan clist (differently from ASH5a) for a literal that yields an atom.
There are four sub-heuristics.

i. Scan the base clauses for a literal with at most one variable. Ground the literal if necessary and select it as the next atom.

ii. Scan the other clauses on clist for a literal with at most one variable. Ground the literal if necessary and select it as the next atom.

iii. Scan the base clauses for any literal. Ground it if necessary and select it as the next atom.

iv. Scan the other clauses on clist for a literal with any number of variables. Ground the literal if necessary and select it as the next atom.

There exist many other possible atom selection heuristics, of course. For example, if at node N there are four two-literal clauses on clist such as A | B, ~A | B, A | ~B, and ~A | ~B, a closed, semantic tree can be constructed by selecting A and B as atoms. To find such a pattern requires considerable computation and it, as well as any other atom selection heuristic that has been explored, has not shown itself to be sufficiently effective to add to the program.

It should be pointed out that ASH1–ASH3 create closed semantic trees at the nodes at which they are successful. In a sense, they are greedy heuristics — heuristics that work well locally but guarantee no overall success. ASH4 also often creates closed semantic trees at nodes at which it is successful. Nodes at which ASH5 selects an atom usually have one child that is a failure node.

**Atom selection heuristic 0 (ASH 0). User selects atom.**
When the d2 and i1 options are used, the user can select atoms. At each node, HERBY first asks the user to select an atom if none has yet been selected. The user cannot pick just any atom. The atom must be able to be obtained by instantiating a literal on clist. The user picks a clause, a literal, and a constant. All variables are assigned the same constant (more about this in Chapter 8).

**Constant selection heuristic (CSH).** If a theorem has only one constant, HERBY always assigns that constant to any literal that it grounds. If there are $n > 1$ constants, HERBY grounds a literal at ply i by assigning it the $(i + 1)$-st constant modulo n. This heuristic needs improvement. At least, constants in the negated conclusion should be assigned more often than others.

## 7.2 Additional Heuristics

HERBY uses several additional heuristics to improve its overall capabilities. They are described in the following sections.

### 7.2.1 List ordering heuristics

There has been an attempt to order lists as they are grown in the optimal order when tested on various theorems. The direction in which lists are scanned has also been optimized over the theorems used to test HERBY.

#### 7.2.1.1 Ordering clauses at each node

HERBY begins by ordering base clauses on clist as follows. The axioms are ordered from longest to shortest based on the number of literals and terms in each clause. Clauses with the most literals are placed at the top of clist. Clauses with an equal number of literals are ordered by the number of terms in the clause. This ordered list is followed by a similarly ordered list of clauses that constitute the negated conclusion. Clauses are ordered at other nodes in the semantic tree in the same way. Generally, short clauses are more useful in constructing a closed semantic tree than long ones, and ordering clauses in this way makes short clauses more accessible.

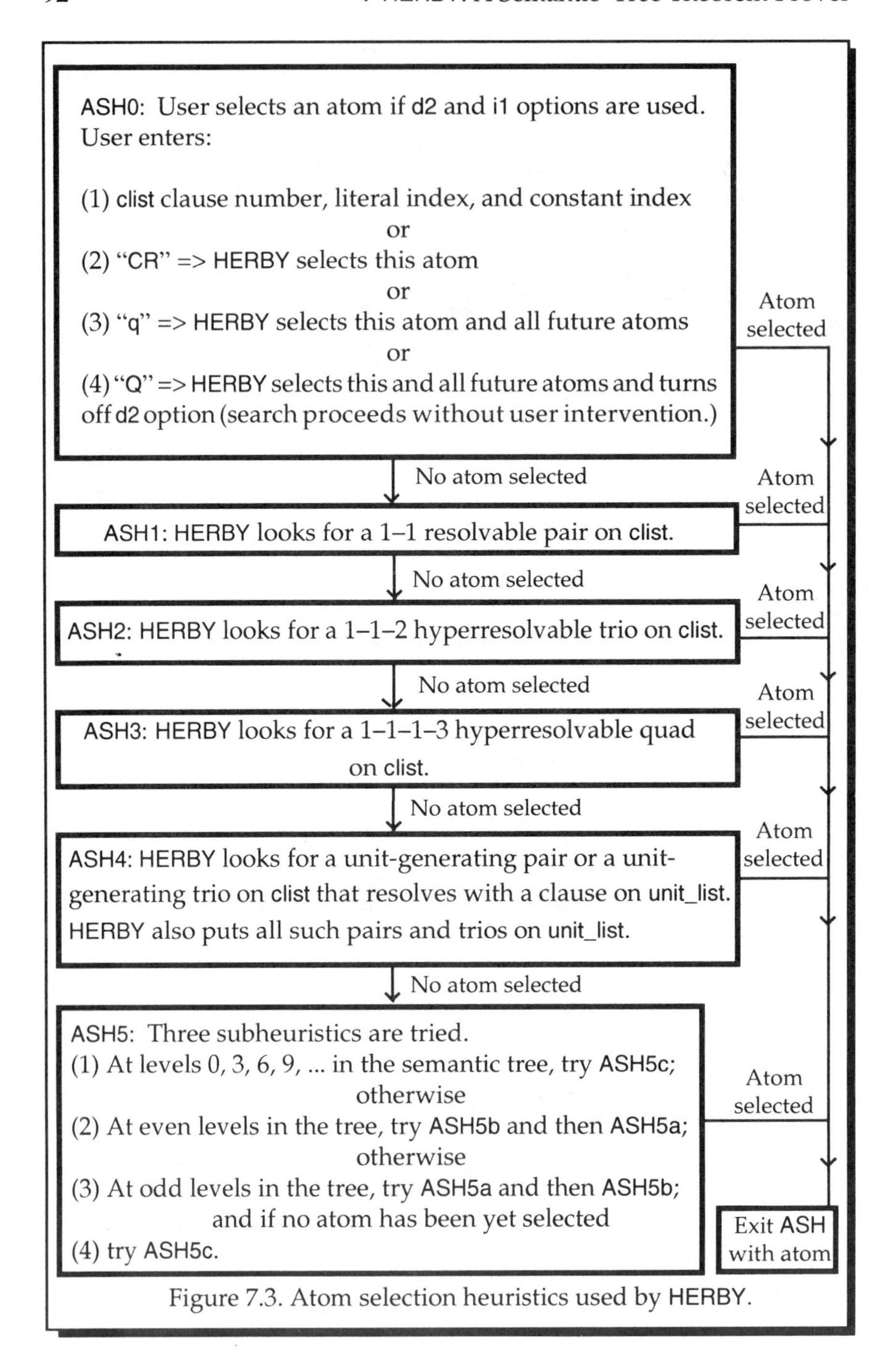


Figure 7.3. Atom selection heuristics used by HERBY.

#### 7.2.1.2 Ordering clauses on unit_list

As unit clauses are found by ASH4, they are placed on unit_list and sorted by length. The shorter units are placed at the top of the list. When ASH4 searches this list to find a clause that contradicts the one currently under consideration, it begins at the top.

### 7.2.2 Preliminary phase (Phase 0): Base clause resolution heuristic

In Phase 0 and before trying to construct a closed semantic tree, HERBY uses the base clause resolution heuristic to create an expanded set of base clause. This heuristic generally makes the atom selection heuristics more effective. HERBY adds up to 25 clauses to the given set of base clauses in the following way. It first resolves the last base clause on clist with all the other base clauses, generating a larger revised set of base clauses. It then resolves this new list with the last base clause and the second to last base clause, generating a still larger set, and so on, until 25 resolvents have been added. Clauses longer than some arbitrary length are not added. Phase 1 is then entered, in which an attempt is made to construct a closed semantic tree.

This is a relatively undesirable heuristic from a theoretical point of view, but it seems to greatly improve HERBY's capabilities. It is one of those heuristics that might be considered a "tuning heuristic," added to HERBY to "tune" it to solve the theorems in the experimental test sets.

### 7.2.3 Heuristic bounding the number of literals in a clause

HERBY limits the number of literals in a resolvent generated by the BCRH to a value of one less than the maximum number of literals in a base clause.

### 7.2.4 Heuristic bounding the number of terms in a literal

HERBY will not generate a clause at a node in the semantic tree if the clause has a literal with more than some number of terms, where this number is related to the maximum number of terms in any base clause. ASH4 may instruct HERBY to increase this limit in order to obtain a locally closed semantic tree.

### 7.2.5 Tree-pruning heuristic

Unlike the other heuristics described thus far, this is really not a heuristic but a 100% safe rule. If HERBY finds that one branch of a subtree rooted at node N fails without the atom rooted there resolving with a clause that caused a failure somewhere in the subtree, then HERBY skips trying to construct the other subtree rooted there. It knows that that subtree is closed. We leave it for the reader to justify this.

## 7.3 Assigning a Hash Code to a Literal and to a Clause

HERBY assigns a number to each literal and each clause on clist and unit_list. This number, called a hash code, makes it easy to determine whether two clauses are identical or whether one clauses s-subsumes another. The details of assigning hash codes to literals and clauses are described in conjunction with the presentation of material on THEO in Sections 9.10.1 and 9.10.2. The two programs use identical algorithms to make this assignment.

## 7.4 The Overall Algorithm

HERBY uses the heuristics described in the previous sections to help construct a closed semantic tree. The overall algorithm follows.

**Algorithm 7.1. The semantic-tree construction procedure of HERBY**

1. Set a time limit for the construction. (One hour is normally used.)
2. Set a bound on the maximum number of literals in a clause and the maximum number of terms in a literal.
3. Phase 0: Add clauses generated by the base clause resolution heuristic to the set of base clauses.
4. Set the depth D of the semantic tree under construction to 0.
5. Phase 1: Construct a semantic tree of depth D.

5a. If all nodes at depth D or less are found to be failure nodes, the construction is complete and the algorithm terminates.

5b. If and whenever a node M at depth D is found not to be a failure node, increase D by one, choose some atom using, in order, ASH0, ASH1, ASH2, ASH3, ASH4, ASH5, and thereafter label all branches at depth D with this atom or its complement. Do not choose an atom that has been previously selected. Continue the construction from M.

5c. If the time limit has been reached, stop.

## 7.5 Obtaining a Resolution–Refutation Proof

Once a closed semantic tree has been constructed, HERBY will go on to obtain a resolution–refutation proof using Algorithm 6.1 described in Section 6.3. HERBY picks the two failure nodes that are siblings and that are the shallowest pair in the semantic tree as the clauses for C1 and C2 in the algorithm.

## Exercises for Chapter 7

7.1. Suggest and discuss other possible atom selection heuristics.

7.2. Discuss the computational complexity of ASH1, ASH2, ASH3, and ASH4.

7.3. Illustrate the semantic tree-pruning heuristic described in Section 7.2.5 with an example.

PRO
THEO

# 8 Using HERBY

This chapter describes how to use HERBY to prove theorems. The input to HERBY is a text file formatted in either its own HERBY/THEO format or in the format of the TPTP Problem Library. This is described in Section 8.1. HERBY saves the results in an output file, as described in Section 8.2. Options that the user can control are described in Section 8.3. User interaction during the search is described in Section 8.4. Examples of user options are presented in Section 8.5. Examples of HERBY's output are given in Sections 8.6 and 8.7. The construction of a closed semantic tree is illustrated in Section 8.6. After finding a closed semantic tree, HERBY can construct a resolution–refutation proof using Algorithm 6.1, as illustrated in Section 8.7.

## 8.1 Proving Theorems with HERBY: The Input File

A theorem that HERBY is asked to prove must be in a text file. It must be formatted in either HERBY/THEO format or in the format used by theorems in the TPTP Problem Library.

In HERBY/THEO format, the character string negated_conclusion must separate the given axioms from the negated conclusion. Each clause can be no longer than a single line. Comments are any characters on a line following a semicolon. The various theorems provided with this package should be examined for more specifics.

For theorems in TPTP Problem Library format, there may be axioms in include files. These files must be in a subdirectory of the directory containing the theorem or located two levels above the theorem. Again, the theorems in the TPTP Problem Library should be examined for specifics.

To prove S39WOS15.THM, in HERBY/THEO format, for example, simply type the command:

```
herby
```

at the system prompt. HERBY will respond:

Enter name of theorem (Type '?' for help):

You should then enter the name of the text file, in this case S39WOS15.THM, including the directory path if the file is not in the active directory. Following the name of the file, you may enter the options you wish HERBY to use, or you may enter a carriage return and let HERBY use its own default settings for all options. Each option consists of a letter followed (with no intervening spaces) by an integer. Options are separated by spaces.

When you enter the carriage return, HERBY will then set out to find a proof. After 3600 seconds have passed, after a time limit established by you has been exceeded and no proof found, or after the tree has reached a depth of 200 levels, HERBY will give up. If and when a proof is found, the atoms used to construct a closed semantic tree are printed on the screen along with some basic related statistics. Everything printed on the screen is also stored on a disk file.

## 8.2 HERBY's Convention on Naming the Output File

The output file is assigned the same name as the input file, except, a ".h" or a ".Hx" suffix either replaces anything following the first period (".") in the name of the input file or is appended to the end of the name of the input file. If HERBY does not find a proof, a .h suffix is added. If a proof is found, a .Hx suffix is added, where x is the time in seconds required to find a solution. For example, if the input file is S39WOS15.THM and a proof is found in 46 seconds, the output file is named S39WOS15.H46. If no proof is found, the results of the search are placed in S39WOS15.h.

## 8.3 The Options Available to the User

There are nine options available to the user when proving a theorem with HERBY. Except for Option 9, they must be entered when entering the name of the theorem. A space separates each option.

OPTION 1: Perform the base clause resolution heuristic (Phase 0) (see Section 7.2.3).

<u>k1</u> Perform the simplification phase before entering Phase 1.

k0 Do not perfom the simplification phase.

OPTION 2: Find a resolution–refutation proof (see Sections 6.3 and 7.4).

r0 Do not find a resolution–refutation proof.

r1 Find a resolution-refutation proof.

OPTION 3: Bound the number of literals in a resolvent in Phase 0 (see

li Section 7.2.2). Eliminate clauses with more than i literals in Phase 0. The default value depends on the maximum number of literals in a clause of the theorem and is determined in the function prepare_clauses in herby.c.

OPTION 4: Bound the number of terms in a literal (see Section 7.2.4).

xi Where i is the maximum number of terms in any literal. The default value depends on the number of terms in the literal of the theorem that is the longest. The default value is determined in the function prepare_clauses in herby.c.

OPTION 5: Set the maximum time given to HERBY to find a proof.

tx Where x is the time in seconds that HERBY is given to prove the theorem. The default value is one hour.

OPTION 6: Print out information as the tree construction progresses.

di If d1, extensive output for detailed observation of how HERBY works. If d2, yet more extensive. The default value is d0.

OPTION 7: Let the user select the atoms.

i0 HERBY selects the atoms.

i1 The user selects the atoms. Used in conjunction with the d2 option.

OPTION 8: Select the TPTP Problem Library format for the input clauses.

tptp If tptp is included, HERBY assumes the theorem is in TPTP Problem Library format.

### 8.3.1 Option to prove a set of theorems

OPTION 9: Prove a set of theorems listed in some file, say, PROVEALL.

batch Tells HERBY that there is a set of theorems in PROVEALL for which proofs are sought. PROVEALL must be proceeded by the redirection symbol, "<." At the system prompt, type herby batch t30 < PROVEALL to tell HERBY to prove every theorem in the file PROVEALL with a time limit of 30 seconds per theorem.

### 8.3.2 Obtaining help by typing "?"

If a "?" appears anywhere on the line naming the theorem, a help menu appears on the screen. After presenting the options available, HERBY again requests the name of a theorem.

## 8.4 User Interaction During the Construction

If options d2 and i1 are used, HERBY will print information about the construction of the semantic tree as it proceeds and will also ask the user to select atoms. The user can then select an instance of any literal that appears on clist. When requested to select an atom, the user enters either two or three integers with a space between each integer. The first integer is the position of the clause on clist, the second integer is the index of the literal, and the third integer is the index of the constant that is to be substituted for all variables in the literal. If the literal has no variables, the third integer is not necessary.

The shortcoming with this procedure is that the user cannot enter just any literal, but it must be a ground instance of a literal on clist, and further, the same constant must be used to replace all variables.

## 8.5 Examples of User Options

Two examples of running HERBY follow. To prove STARK103.THM with a time limit of 30 seconds, with a maximum of eight terms per literal, and with extensive output describing the construction process, type:

```
STARK103.THM  t30 x8 d2
```

To prove GEO001-1.p in the directory GEO in the TPTP Problem Library with the user able to select the atoms, type:

```
GEO001-1.p tptp d2 i1
```

## 8.6 The Printout Produced by **HERBY**

STARK103.THM serves as a good example. The printout produced by HERBY when using the option d2 follows. Comments have been added. The bold type was produced by HERBY, and the underlined type by the user (this occurs only twice). Material in square brackets represents comments added to explain the printout. At the system prompt, type:

```
HERBY

Enter name of theorem (type '?' for help): STARK103.THM d2

STARK103.THM

Predicates: S M E UN   Functions: B A : . G F
```

[HERBY lists the predicates and functions found in the clauses of the theorem. The number of arguments for each predicate and each function is not indicated but can be determined by examining the clauses. The functions are separated into three groups. The first group contains constants in the negated conclusion — **B** and **A** in this case. The second group, listed after the colon and before the period, contains the other constants (which in this case is none). The third group, listed after the period, contains the functions in the theorem, **G** and **F**.]

```
EQ:
ESAF:                    ESAP:
```

[HERBY lists axioms that are equality substitution axioms for functions and for predicates. For this theorem, there are not any. Note that clauses are printed without parentheses or logical " | " symbols. Both can normally be inferred by carefully examining the base clauses.]

```
27 clauses are added by the BCR Heuristic:

 15: (14a,6c)   ~SBA ~SAB
 16: (13a,9c)   ~MxA MxB
 17: (13a,7d)   ~MxB MxA
 18: (12a,17b)  ~MGxAyy UNxAy ~MGxAyB
 19: (12b,17b)  ~MGxyAy UNxyA ~MGxyAB
 20: (12a,16b)  ~MGxByy UNxBy ~MGxByA
 21: (12b,16b)  ~MGxyBy UNxyB ~MGxyBA
 22: (11a,17b)  ~MGAxyy UNAxy ~MGAxyB
 23: (11b,17b)  ~MGxyAx UNxyA ~MGxyAB
 24: (11a,16b)  ~MGBxyy UNBxy ~MGBxyA
 25: (11b,16b)  ~MGxyBx UNxyB ~MGxyBA
 26: (9a,17b)   Mxy ~UNzAy ~MxB
 27: (9b,17a)   ~Mxy ~UNzyB MxA
```

```
 28: (9a,16b)  Mxy ~UNzBy ~MxA
 29: (9b,16a)  ~Mxy ~UNzyA MxB
 30: (9b,3b)   ~MFxyz ~UNuzy Sxy
 31: (9a,2b)   MFxyz ~UNuxz Sxy
 32: (13a,31b) MFAxB SAx
 33: (13a,30b) ~MFxBA SxB
 34: (9b,33a)  ~MFxBy ~UNzyA SxB
 35: (9a,32a)  MFAxy ~UNzBy SAx
 36: (8b,33a)  ~MFxBy ~UNyzA SxB
 37: (8a,32a)  MFAxy ~UNBzy SAx
 38: (8a,17b)  Mxy ~UNAzy ~MxB
 39: (8b,17a)  ~Mxy ~UNyzB MxA
 40: (8a,16b)  Mxy ~UNBzy ~MxA
 41: (8b,16a)  ~Mxy ~UNyzA MxB

NEXTC=41 TIME=3600 XARS=25
```

[HERBY is now ready to try to construct a closed semantic tree using an extended set of 41 base clauses, a time limit of 3600 seconds, and a maximum of 25 terms in a literal.]

```
GENERATE ATOM   1: ~SBA   H5aiv.15   T0   N1   C0   U0
```

[At the root of the semantic tree, HERBY uses ASH5aiv to select an atom; that heuristic found a grounded two-literal clause as the 15th clause on clist. **T0** denotes the amount of time that passed (0 seconds); **N1** indicates that one node has been constructed so far. **C0** gives the number of nonbase clauses on clist (none at this time), and **U0** indicates that there are no unit clauses on unit_list.]

```
Branch on atom:   1: SBA to node 1

GENERATE CLAUSES AT NODE 1
 42# (1a,6a) ~SBA#1  ~SAB EBA
 43# (1a,6b) ~SAB ~SBA#1  EAB
 44: (1a,15a)~SBA#1  ~SAB
GENERATED 3 CLAUSES
PATH: 1
```

[A path in the semantic tree is described by a sequence of **1**s and **2**s, where a **1** corresponds to a left-hand branch and a **2** corresponds to a right-hand branch. The path to node **1**, the left-hand branch of the root, is thus "**1**."]

```
GENERATE ATOM USING ASH4a:
```

```
C1,C2 resolve to C3

C1:  44: (1a,15a)   ~SBA#1   ~SAB
C2:   3: Sxy ~MFxyy
C3:  45: (44a,3a)   ~MFABB

  2: ~MFABB    H4a
  3: ~SAB    -H4   T1   N2   C3   U6
```

[ASH4 found a 1–2 unit-generating pair, clauses **44** and **3** on clist and the unit clause **~MFABB** on unit_list that resolved to yield the NULL clause and providing atoms **2** and **3** as shown. The first literal of clause **44**, **~SBA**, was resolved away by the first atom, as the **#1** indicates. One second has passed, two nodes examined, three clauses are on clist, and six unit clauses are on unit_list.]

```
Branch on atom:    2: MFABB to node 2

GENERATE CLAUSES AT NODE 2
 45: (2a,1b)    ~SBx ~MFABB#2   MFABx
 46: (2a,3b)    SAB ~MFABB#2
 47: (2a,7a)    ~MFABB#2   MFABx MFABy ~UNxyB
 48: (2a,8a)    ~MFABB#2   MFABx ~UNByx
 49: (2a,9a)    ~MFABB#2   MFABx ~UNyBx
 50: (2a,17a)   ~MFABB#2   MFABA
 51: (2a,26c)   MFABx ~UNyAx ~MFABB#2
 52: (2a,38c)   MFABx ~UNAyx ~MFABB#2
GENERATED 8 CLAUSES

PATH: 11

Branch on atom:    3: SAB to node 3
```

[Recall that when the second atom was generated, so was the third, and thus no atom is generated here.]

```
GENERATE CLAUSES AT NODE
PATH: 111 FAILS:   56: (3a,15b)   ~SBA#1   ~SAB#3
```

{Node **3** is a failure node, failing because atom **1** resolves with the first literal of clause **15** and atom **3** with the second literal to yield clause **56**, the NULL clause. Note that it is numbered as the 56th clause. When a failure clause is generated at a node, the other generated clauses are not listed. There were evidently three other clauses generated at this node, clauses **53–55**, before the NULL clause was generated.]

```
Branch on atom:    3: ~SAB to node 4
```

```
GENERATE CLAUSES AT NODE 4

PATH: 112 FAILS:   59: (3a,3a)    SAB#3   ~MFABB#2
PATH: 11 FAILS:
```

[Because `PATH 111` and `112` both fail, so does `PATH 11`.]

```
Branch on atom:    2: ~MFABB to node 5

GENERATE CLAUSES AT NODE 5
 45: (2a,1c)    ~SxB ~MFABx MFABB#2
 46: (2a,7b)    ~MFABx MFABB#2  MFABy ~UNByx
 47: (2a,7c)    ~MFABx MFABy MFABB#2  ~UNyBx
 48: (2a,8b)    ~MFABx MFABB#2  ~UNxyB
 49: (2a,9b)    ~MFABx MFABB#2  ~UNyxB
 50: (2a,16b)   ~MFABA MFABB#2
 51: (2a,29c)   ~MFABx ~UNyxA MFABB#2
 52# (2a,31a)   MFABB#2  ~UNxAB SAB
 53: (2a,32a)   MFABB#2  SAB
 54: (2a,41c)   ~MFABx ~UNxyA MFABB#2
 55: (2a,7c)    ~MFABx MFABB#2  MFABB#2  ~UNBBx
GENERATED 11 CLAUSES

Branch on atom:    3: SAB to node 6
GENERATE CLAUSES AT NODE 6
PATH: 121 FAILS:  59: (3a,15b)  ~SBA#1  ~SAB#3

Branch on atom:    3: ~SAB to node 7

GENERATE CLAUSES AT NODE 7
PATH: 122 FAILS:  62: (3a,32b)  MFABB#2  SAB#3
PATH: 12 FAILS:
PATH: 1 FAILS:
```

[A closed subtree of the left-hand child of the root node has been constructed; now an attempt is made to construct one for the right-hand child.]

```
Branch on atom:    1: ~SBA to node 8

GENERATE CLAUSES AT NODE 8
 42: (1a,2a)    SBA#1  MFBAB
 43: (1a,3a)    SBA#1  ~MFBAA
 44: (1a,5a)    SBA#1  ~EAB
 45: (1a,30c)   ~MFBAx ~UNyxA SBA#1
 46: (1a,31c)   MFBAx ~UNyBx SBA#1
```

```
GENERATED 5 CLAUSES

Branch on atom:   2: MFABB to node 9

GENERATE CLAUSES AT NODE 9
 47: (2a,1b)    ~SBx ~MFABB#2  MFABx
 48: (2a,3b)    SAB ~MFABB#2
 49: (2a,7a)    ~MFABB#2  MFABx MFABy ~UNxyB
 50: (2a,8a)    ~MFABB#2  MFABx ~UNByx
 51: (2a,9a)    ~MFABB#2  MFABx ~UNyBx
 52: (2a,17a)   ~MFABB#2  MFABA
 53: (2a,26c)   MFABx ~UNyAx ~MFABB#2
 54: (2a,38c)   MFABx ~UNAyx ~MFABB#2
GENERATED 8 CLAUSES

Branch on atom:   3: SAB to node 10

GENERATE CLAUSES AT NODE 10
GENERATED 0 CLAUSES

PATH: 211

GENERATE ATOM C1,C2 resolve to C4, which resolves with C3

C1:  43: (1a,3a)    SBA#1  ~MFBAA
C2:  17: (13a,7d)   ~MxB MxA
C3:  42: (1a,2a)    SBA#1  MFBAB
C4:  55: (43a,17b) ~MFBAB
The atom generated is: 56: ~MFBAA

   4: MFBAB   H2
   5: ~MFBAA   -H2  T5  N11  C13  U6
```

[ASH2 selected the fourth and fifth atoms.]

```
Branch on atom:   4: ~MFBAB to node 11

GENERATE CLAUSES AT NODE 11
PATH: 2111 FAILS:  62: (4a,2b)   SBA#1  MFBAB#4

Branch on atom:   4: MFBAB to node 12

GENERATE CLAUSES AT NODE 12
 55: (4a,1b)    ~SBx ~MFBAB#4  MFBAx
 56: (4a,7a)    ~MFBAB#4  MFBAx MFBAy ~UNxyB
 57: (4a,8a)    ~MFBAB#4  MFBAx ~UNByx
```

```
 58: (4a,17a)  ~MFBAB#4  MFBAA
 59: (4a,26c)  MFBAx ~UNyAx ~MFBAB#4
 60# (4a,30a)  ~MFBAB#4  ~UNxBA SBA
 61: (4a,38c)  MFBAx ~UNAyx ~MFBAB#4
 62: (4a,30a)  ~MFBAB#4  ~UNxBA SBA#1
GENERATED 8 CLAUSES

PATH: 2112

Branch on atom:   5: MFBAA to node 13

GENERATE CLAUSES AT NODE 13

PATH: 21121  FAILS:  66: (5a,3b)   SBA#1  ~MFBAA#5

Branch on atom:   5: ~MFBAA to node 14

GENERATE CLAUSES AT NODE 14
PATH: 21122  FAILS:  73: (5a,17b)  ~MFBAB#4  MFBAA#5
PATH: 2112 FAILS:
PATH: 211 FAILS:
PATH: 21 FAILS:
PATH: 2 FAILS:
PATH:  FAILS:

TIME:7 (NODES: 15, ATOMS: 5)
** Proof Found! **
```

[A closed semantic tree has been constructed with 15 nodes and 5 atoms. Clauses placed on unit_list follow, then the theorem, then the atoms, and then a bit of information about the theorem.]

```
UNIT LIST
 45: ~SAB
 45: ~EBA
 45: ~EAB
 46: MFABB
 46: ~MFABA
 45: ~MFABB

Given axioms:
  1: ~Sxy ~Mzx Mzy
  2: Sxy MFxyx
  3: Sxy ~MFxyy
  4: Sxy ~Exy
  5: Sxy ~Eyx
```

```
  6: ~Sxy ~Syx Exy
  7: ~Mxy Mxz Mxu ~UNzuy
  8: ~Mxy Mxz ~UNyuz
  9: ~Mxy Mxz ~UNuyz
 10: MGxyzz MGxyzx MGxyzy UNxyz
 11: ~MGxyzx ~MGxyzz UNxyz
 12: ~MGxyzy ~MGxyzz UNxyz
 13: UNAAB

Negated conclusion:
 14: ~EBA

Atoms used to construct closed semantic tree:
  1: ~SBA
  2: MFABB
  3: SAB
  4: MFBAB
  5: ~MFBAA

Total Time: 0 sec
BASE CLAUSES: 14    REVISED SET OF BASE CLAUSES: 41
Nodes: 15  ATOMS: 5
```

## 8.7 A Second Example: The Printout Produced Using the r1 Option

With the r1 option, HERBY first tries to construct a closed semantic tree, and, if successful, then goes on to construct a resolution–refutation proof. When deciding which two failure nodes in the semantic tree to resolve together, HERBY always picks two nodes that are at the shallowest level in the tree. If there is more than one such pair, HERBY picks the first pair found.

STARK103.THM serves as a good example to illustrate how HERBY finds a resolution–refutation proof. The printout that follows was produced by HERBY; comments have been added in square brackets by the author. The bold type was produced by HERBY; the underlined type by the user.

**Enter name of theorem (type '?' for help):** STARK103.THM r1

[The "r1" option tells HERBY to construct a closed semantic tree and then find a resolution–refutation proof.]

**../t/theorems/STARK103.THM**

```
Predicates: S M E UN
Functions:  B A : . G F
EQ:
ESAF:                ESAP:

NEXTC=41 TIME=3600 XARS=25
  1: ~SBA      H5aiv.15  T0  N1  C0  U0
  2: ~MFABB    H4a
  3: ~SAB      -H4  T0  N2  C3  U6
  4: MFBAB     H2
  5: ~MFBAA    -H2  T0  N11  C13  U6

TIME:0 (NODES: 15, ATOMS: 5)
** Proof Found! **
```

[The same proof is obtained as the one in the previous section. The first resolvent of the proof will now be calculated.]

```
RESOLVE CLAUSES:
 15: (14a,6c)  ~SBA ~SAB
   3: Sxy ~MFxyy
RESOLVE ON ATOM 3:
   3: ~SAB
BINARY RESOLVENT IS:
 42: (15b,3a)  ~SBA ~MFABB

NEXTC=42 TIME=3600 XARS=25

TIME:0 (NODES: 13, ATOMS: 5)
** Proof Found! **
```

[This was the second time that a closed semantic tree was constructed, but this time with only 13 nodes.]

```
RESOLVE CLAUSES:
 15: (14a,6c)  ~SBA ~SAB
 32: (13a,31b) MFAxB SAx
RESOLVE ON ATOM 3:
   3: ~SAB
BINARY RESOLVENT IS:
 43: (15b,32b) ~SBA MFABB
```

[This is the second resolvent of the proof.]

```
NEXTC=43 TIME=3600 XARS=25

TIME:0 (NODES: 11, ATOMS: 5)
** Proof Found! **
```

[This was the third time a closed semantic tree was constructed, but this time with only 11 nodes.]

```
RESOLVE CLAUSES:
 42: (15b,3a)   ~SBA ~MFABB
 43: (15b,32b) ~SBA MFABB
RESOLVE ON ATOM 2:
   2: ~MFABB
BINARY RESOLVENT IS:
 44: (42b,43b) ~SBA ~SBA      [This is the third resolvent of the proof.]
FACTOR RESOLVENT:
 45: (44ab)     ~SBA
```

[The third resolvent was factored and clause **44** was replaced by clause **45**. (see step 1f in Algorithm 6.1.) HERBY actually does not physically replace clause **44** with clause **45**, but uses one bit in the representation of clause **44** to indicate that it is "**DEAD**" and thus never used.]

```
NEXTC=45 TIME=3600 XARS=25

TIME:0 (NODES: 9, ATOMS: 5)
** Proof Found! **

RESOLVE CLAUSES:
   3: Sxy ~MFxyy
 17: (13a,7d)  ~MxB MxA
RESOLVE ON ATOM 5:
   5: ~MFBAA
BINARY RESOLVENT IS:
 46: (3b,17b)  SxA ~MFxAB     [This is the fourth resolvent of the proof.]

NEXTC=46 TIME=3600 XARS=25

TIME:0 (NODES: 7, ATOMS: 4)
** Proof Found! **

RESOLVE CLAUSES:
   2: Sxy MFxyx
 46: (3b,17b)   SxA ~MFxAB
RESOLVE ON ATOM 4:
  4: MFBAB
BINARY RESOLVENT IS:          [This is the fifth resolvent of the proof.]
 47: (2b,46b)   SBA SBA
FACTOR RESOLVENT:
 48: (47ab)    SBA                  [This resolvent must be factored.]
```

```
NEXTC=48 TIME=3600 XARS=25

TIME:0 (NODES: 3, ATOMS: 1)
** Proof Found! **

RESOLVE CLAUSES:
 45: (44ab)    ~SBA
 48: (47ab)    SBA
RESOLVE ON ATOM 1:
   1: ~SBA
BINARY RESOLVENT IS:
 49: (45a,48a) []
```

[The contradiction is generated.]

```
Given axioms:
  1: ~Sxy ~Mzx Mzy
  2: Sxy MFxyx
       .
       .
       .
 13: UNAAB

Negated conclusion:
 14: ~EBA
Atoms used to construct closed semantic tree:
  1: ~SBA
  2: ~MFABB
  3: ~SAB
  4: MFBAB
  5: ~MFBAA

Resolution clauses:
 15: (14a,6c)  ~SBA ~SAB
 16: (13a,9c)  ~MxA MxB
 17: (13a,7d)  ~MxB MxA
       .
       .
       .
 39: (8b,17a)  ~Mxy ~UNyzB MxA
 40: (8a,16b)  Mxy ~UNBzy ~MxA
 41: (8b,16a)  ~Mxy ~UNyzA MxB
```

[The resolution–refutation proof, all together, is:]

```
Proof:
 42: (15b,3a)     ~SBA ~MFABB
 43: (15b,32b)   ~SBA MFABB
 44# (42b,43b)   ~SBA ~SBA
 45: (44ab)        ~SBA
 46: (3b,17b)     SxA ~MFxAB
 47# (2b,46b)     SBA SBA
 48: (47ab)        SBA
 49: (45a,48a)    []

Total Time: 0 sec
BASE CLAUSES: 14   REVISED SET OF BASE CLAUSES: 41
Nodes: 3  ATOMS: 5
```

## Exercises for Chapter 8

8.1. Use HERBY to find a closed semantic tree for STARK108.THM. Draw the tree, labeling the branches with the atoms and the terminal nodes with the clauses that generate the NULL clause. Was the tree-pruning heuristic described in Section 7.2.5 used? If so, show where in the tree.

8.2. Use HERBY to find a resolution–refutation proof of STARK118.THM.

8.3. How many theorems in the Stickel test set can HERBY solve in 10 seconds? In 30 seconds? In 60 seconds? In 300 seconds? In 1200 seconds?

8.4. HERBY is unable is solve S44WOS20.THM in any reasonable amount of time. See if you can construct a closed semantic tree in only several minutes by using HERBY's options i1 and d2, which permit you to select atoms.

8.5. [Project] Modify ASH0 to permit the user to enter arbitrary atoms.

THEO

# 9 THEO: A Resolution–Refutation Theorem Prover

This chapter describes the procedure used by THEO when attempting to find a proof. THEO passes through four phases during the process. A preliminary phase, Phase 0, attempts to simplify the given base clauses. In Phase 1, a search for a proof is carried out. This phase ends when some main line is found; that is, when THEO knows that it has a proof. If necessary, additional lines are found in the next phase, Phase 2. In Phase 3, the proof is checked for correctness and then printed.

In Phase 0, before beginning to search for a proof, THEO attempts to simplify the base clauses. The user can instruct THEO to skip this phase. The simplification procedure is carried out by the function eliminate as described in Section 9.24.

In Phase 1, THEO attempts to find a linear proof by carrying out an iteratively deepening depth-first search, as described in Section 9.1. THEO can restrict the search to a linear-merge proof, a linear-nc proof, or a linear-merge-nc proof, as described in Sections 9.2–9.4. The default strategy is to search for a linear-merge proof. In Section 9.5, the extended search strategy is described. This strategy, normally used by THEO, tends to focus the search. Sections 9.6–9.8 describe three heuristics used to place upper bounds on the number of literals, terms, and variables in an inference. The user can override these limits, although the program's data structures impose absolute upper bounds. Section 9.9 describes how THEO orders inferences for search at each node.

While searching for a linear proof, THEO may find a more complex proof. If it does, it enters Phase 2, where it attempts to put together the pieces; otherwise, it goes directly to Phase 3. THEO uses a large hash table to assist in obtaining proofs that are more complex than linear. In Sections 9.10.1 and 9.10.2, algorithms for assigning a hash code to a literal and for assigning a hash code to a clause are described. Entering clauses in a large hash table is described in Section 9.10.3. In Section 9.11, the use of the hash table to reduce the size of the search space by eliminating previously hashed inferences is described. Unit clauses entered into the hash table are used when generating unit hash table resolutions, as described in Section 9.12. The hashing of instants and variants of unit clauses is described in Section 9.13. The hashing of still other unit clauses is described in Section 9.14. Sections 9.15-9.22 describe seven search strategies, none of which the user has any control over. Section 9.23 discusses the use of the array, cl, in determining a proof. This array is used to store lines of the proof as they are found.

In Phase 3, THEO verifies the proof found in earlier phases. If a hashing error had erroneously yielded a proof, it would be detected here.

Section 9.25 summarizes the search heuristics.

## 9.1 Iteratively Deepening Depth-First Search and Linear Proofs

When THEO attempts to find a proof, it carries out a sequence of deeper and deeper searches, or **iterations**, into a **resolution–refutation search tree** of inferences. The **nodes** of the tree correspond to clauses, and the **branches** correspond to inferences performed. The **root** of the search tree corresponds to the set of base clauses. Branches leading from one node to another correspond to inferences that can be performed on the clauses at which the branch is rooted. Each **nonroot** node corresponds to the clause generated by the inference on the branch leading to it.

On the first iteration, a search for a linear proof of length one is carried out, on the second, a search for a linear proof of length two is carried out, and so on. Finding a linear proof of length one on the first iteration occurs if two base clauses resolve together to yield the NULL clause. If the NULL clause is generated, the procedure terminates with a proof. Finding a linear proof of length two on the second iteration involves resolving together all pairs of base clauses and factoring individual base clauses and then resolving each of the resulting first-level inferences with some base clause. Again, if the NULL clause is generated, the procedure terminates with a proof of length two. Otherwise, a third iteration is carried out, this time looking for a proof of length three.

In Chapter 6, it was shown that if a theorem has a proof, it has a linear proof. Because the nth iteration of an iteratively deepening depth-first search will find a linear proof having n inferences if one exists, and because for any n, eventually the nth iteration will be carried out, this search strategy is complete; it will always find a proof if one exists.

Depth-first search — as contrasted with breadth-first search — is a natural way to look for a linear proof, and moreover, carrying out an iteratively deepening depth-first search solves the problem of not knowing a priori how deep to set the search bound when the length of the proof is unknown. The time taken by the first few iterations is generally only a small fraction of the total search time.

At each node in the search tree, inferences are generated as shown in the block of pseudo-C code in Figure 9.2. The code assumes the base clauses are $BC_1, BC_2, ..., BC_M$ and the inferences on the path to and including some node K at level N in the tree are $I_1, I_2, ..., I_N$.

Figure 9.3 shows a printout of the search that THEO carries out for A.THM shown in Figure 9.1. On the first iteration, when the depth of search is one, three clauses are generated and no proof is found. On the second iteration, nine clauses are generated and no proof is found. On the third iteration, the NULL clause is the fourth clause generated. A total of 16 clauses are generated during the three iterations.

```
Axioms
1  P(x) | ~Q(x)
2  Q(a) | Q(b)

Negated conclusion
3  ~P(x)
```

Figure 9.1. A.THM.

```
If(K is the root node) {
    For (i = M ; i ≥ 1 ;  i--) {
        Form factors of BC_i;
        For(j = i - 1 ; j ≥ 1 ;  j--) {
            Form resolvents of BC_i and BC_j;}}}
Else if(K is not the root node) {
    Form factors of clause I_N;
    For(i = M ;  i ≥ 1 ;  i--) {
        Form resolvents of BC_i and  inference I_N;}
    For(i = N - 2;  i > 0 ;  i--) {
        Form resolvents of inference I_i and  inference I_N;}}
```

Figure 9.2. Pseudo-C code for generating inferences at some node K at depth N in the search of a linear proof of a theorem with M base clauses, $BC_1, BC_2, ..., BC_M$, and with inferences $I_1, I_2, ..., I_{N-1}$ on the path to node K.

On the first iteration with the maximum search depth set to one, all binary resolvents and all binary factors of the base clauses are generated:

4: (3a,1a) ~Q(x)
4: (2b,1b) P(b) | Q(a)
4: (2a,1b) P(a) | Q(b)

Because the first iteration fails to find the NULL clause, a second iteration that generates all sequences of inferences of length two is carried out:

4: (3a,1a) ~Q(x)
5: (4a,2b) Q(a)
5: (4a,2a) Q(b)

4: (2b,1b) P(b) | Q(a)
5: (4a,3a) Q(a)
5: (4b,1b) P(a) | P(b)

4: (2a,1b) P(a) | Q(b)
5: (4a,3a) Q(b)
5: (4b,1b) P(a) | P(b)

The second iteration fails to find the NULL clause also. A third iteration is thus necessary, and this one yields a linear proof:

4: (3a,1a) ~Q(x)
5: (4a,2b) Q(a)
6: (5a,1b) ~P(a)
6: (5a,4a) Ø

Figure 9.3. Three iterations carried out by THEO when searching for a linear proof of A.THM.

## 9.2 Searching for a Linear-Merge Proof

The user can instruct THEO to restrict or not to restrict its search to finding a linear-merge proof. Normally, THEO so restricts the search. When searching for a linear-merge proof, at each node in the search tree it is necessary to generate only the binary resolvents of the clause or clauses (in the case of the root node) at that node with merge clauses or base clauses at an ancestor node. Factoring is performed on the clause or clauses at that node, too. Imposing this additional restriction on the search for a linear proof of A.THM yields the four search trees shown in Figure 9.4. The first two trees are identical to those searched in Figure 9.3, where only a linear proof is sought. However, this time no proof is found on the third ieration; it was found on the third iteration in Figure 9.3 when clauses 4 and 5 at the third level in the search tree were resolved together. Because clause 4 is not a merge clause, this time it is not resolved with clause 5. A proof of length four is found this time. For this theorem, looking for a linear-merge proof increases the size of the search space and even increases the length of the proof,

The first iteration with maximum search depth set to one (all inferences of the base clauses):

```
4: (3a,1a)  ~Q(x)
4: (2b,1b)  P(b) | Q(a)
4: (2a,1b)  P(a) | Q(b)
```

Because the first iteration fails to find the NULL clause, a second iteration that searches all sequences of inferences of length two is carried out:

```
4: (3a,1a)  ~Q(x)          5: (4a,2b)  Q(a)
                           5: (4a,2a)  Q(b)
4: (2b,1b)  P(b) | Q(a)    5: (4a,3a)  Q(a)
                           5: (4b,1b)  P(a) | P(b)
4: (2a,1b)  P(a) | Q(b)    5: (4a,3a)  Q(b)
                           5: (4b,1b)  P(a) | P(b)
```

The second iteration fails to find the NULL clause also. A third iteration is thus necessary.

```
4: (3a,1a)  ~Q(x)          5: (4a,2b)  Q(a)          6: (5a,1b)  P(a)
                           5: (4a,2a)  Q(b)          6: (5a,1b)  P(b)
4: (2b,1b)  P(b) | Q(a)    5: (4a,3a)  Q(a)          6: (5a,1b)  P(a)
                           5: (4b,1b)  P(a) | P(b)   6: (5b,3a)  P(a)
                                                     6: (5a,3a)  P(b)
4: (2a,1b)  P(a) | Q(b)    5: (4a,3a)  Q(b)          6: (5a,1b)  P(b)
                           5: (4b,1b)  P(a) | P(b)   6: (5b,3a)  P(a)
                                                     6: (5a,3a)  P(b)
```

The third iteration fails to find the NULL clause. Note that because 4: (3a,1a) ~Q(x) is not a merge clause, it was not resolved with 5: (4a,2b) Q(a) as was done in the tree generated in Figure 9.2. The fourth iteration does:

```
4: (3a,1a)  ~Q(x)  5: (4a,2b)  Q(a)  6: (5a,1b)  P(a)  7: (6a,3a)  Ø
```

Figure 9.4. Four iterations carried out by THEO when searching for a linear-merge proof of A.THM.

making the strategy look somewhat ineffective. However, even though the strategy can result in longer proofs, as happened here, it usually does reduce the search space, yielding a proof in less time than a less-restricted search for a linear proof.

In Chapter 6, it was shown that if a theorem has a proof, it has a linear-merge proof of some length, say n. The nth iteration of an iteratively deepening depth-first search will find a linear-merge proof having n inferences if one exists. Because for any n, the nth iteration eventually will be carried out, this search strategy is complete.

## 9.3 Searching for a Linear-nc Proof

The user can instruct THEO to restrict or not to restrict its search to finding a linear-nc proof. Normally, THEO does not so restrict the search. If it does, inferences at the first level must be descendants of clauses that constitute the negated conclusion. There is only one such clause in A.THM, and when restricting the search for a proof of this theorem to a linear-nc proof, the resulting search is as shown in Figure 9.5.

In Chapter 6, it was shown that if a theorem has a proof, it has a linear-nc proof. The nth iteration of an iteratively deepening depth-first search will

The first iteration with the maximum search depth set to one:

```
4: (3a,1a)  ~Q(x)
```

Because the first iteration fails to find the NULL clause, a second iteration that searches all sequences of inferences of length two is carried out next:

```
4: (3a,1a)  ~Q(x)                  5: (4a,2b)  Q(a)
                                   5: (4a,2a)  Q(b)
```

The second iteration fails to find the NULL clause also. A third iteration is thus necessary.

```
4: (3a,1a)  ~Q(x)     5: (4a,2b)  Q(a)      6: (5a,4a)  Ø
```

Figure 9.5. Three iterations carried out by THEO when searching for a linear-nc proof to A.THM.

find a linear-nc proof having n inferences if one exists. For any n, eventually the nth iteration will be carried out, and thus this search strategy is complete.

## 9.4 Searching for a Linear-Merge-nc Proof

By default, THEO searches for a linear-merge proof but not a linear-nc proof. If it combined both strategies when proving A.THM, the search trees shown in Figure 9.6 would have been generated. Again, a fourth iteration is necessary. It is not clear whether searching for a linear-merge-nc proof is more or less effective than searching only for a linear-merge proof.

In Chapter 6, it was shown that if a theorem has a proof, it has a linear-merge-nc proof. If the linear-merge-nc proof is of length n, it will be found on the nth iteration. For any n, eventually the nth iteration will be carried out, and thus this strategy is complete.

The first iteration with the maximum search depth set to one:

```
4: (3a,1a) ~Q(x)
```

Because the first iteration fails to find the NULL clause, a second iteration that searches all sequences of inferences of length two is carried out:

```
4: (3a,1a)  ~Q(x)      5: (4a,2b) Q(a)
                       5: (4a,2a) Q(b)
```

The second iteration fails to find the NULL clause also. A third iteration is thus necessary.

```
4: (3a,1a) ~Q(x)       5: (4a,2b)  Q(a)          6: (5a,1b)  P(a)
                       5: (4a,2a)  Q(b)          6: (5a,1b)  P(b)
```

The third iteration fails to find the NULL clause. A fourth iteration is necessary.

```
4: (3a,1a)  ~Q(x)   5: (4a,2b)  Q(a)   6: (5a,1b)  P(a)    7: (6a,3a)  Ø
```

Figure 9.6. Four iterations carried out by THEO when searching for a linear-merge-nc proof of A.THM.

## 9.5 The Extended Search Strategy

By default, THEO extends the search beyond level n of a tree on the nth iteration under certain conditions; the user can override this strategy. Extended search is carried out on some clause C if the extended search option is selected and if C is at level n or deeper and if it can be factored or if it can be resolved with a unit clause C', which is one of its ancestors. Clauses generated as a result of this strategy are said to be in the **extended search region**. Clauses at level n on the nth iteration are said to be on the **search horizon**.

Extended search as carried out by THEO is analogous to the extended search carried out by chess programs when looking for highly forced lines of play deep in their search trees. These lines usually involve captures and checks. When the extended search strategy is combined with a search for a linear-merge-nc proof of A.THM, the trees shown in Figure 9.7 result. The proof is found after only two iterations and after generating only four inferences.

The extended search strategy maintains completeness in the sense defined for the previous search strategies presented thus far in this chapter, but it goes further. If a theorem has a linear proof of length n, it will be found on the kth iteration, where $k \leq n$, of an iteratively deepening depth-first search. For A.THM, a proof of length three was found on the second iteration.

The first iteration with the maximum search depth set to one:

4: (3a,1a) ~Q(x)

Because the first iteration fails to find the NULL clause, a second iteration that searches all sequences of inferences of length two is carried out next:

4: (3a,1a) ~Q(x)   5: (4a,2b) Q(a)   6: (5a,4a) Ø

The proof is found on the first line searched! A search for a linear-merge-nc proof using the extended search strategy yielded a proof after searching only four inferences.

Figure 9.7. Two iterations carried out by THEO when searching for a linear-merge-nc proof of A.THM using the extended search strategy.

## 9.6 Bounding the Number of Literals in a Clause

If two clauses C1 and C2 are resolved together, where C1 has i literals and C2 has j literals, the resolvent may have as many as i + j – 2 literals. Thus, if the number of literals in C1 or C2 is three or greater, their resolvent may be longer than either of them. In general, resolving long clauses together generates even longer clauses. Because the objective is to generate the NULL clause, the creation of long clauses seems counterproductive. Let LITmaxbase denote the maximum number of literals in any base clause. THEO normally limits the number of literals in a resolvent, denoted o_maxlit, as follows. Before a search begins, it determines LITmaxbase and sets o_maxlit to LITmaxbase – 1; a minimum value for o_maxlit is three. The user can override this value for o_maxlit but it cannot exceed 20. During the search for a proof, a resolvent is discarded if it has more than o_maxlit literals. It is left for the reader in Exercise 9.5 to show that this strategy sacrifices completeness.

## 9.7 Bounding the Number of Terms in a Literal

As inferences are formed, the number of terms in their literals is often greater than the number of terms in their parent clauses. Because most proofs seem not to have that many more terms in the literals of the inferences than in the original base clauses, THEO initially places a limit on the size of terms and stores this bound in o_maxterm. It does this using a complex formula that tries to optimize the number of problems solved in the TPTP Problem Library. When inferences are generated, those with more than o_maxterm are eliminated. This heuristic sacrifices completeness. The user, however, can set the value of o_maxterm to any value from 2 to 199 before beginning the search for a proof.

## 9.8 Bounding the Number of Different Variables in an Inference

THEO places a bound on the number of different variables that may appear in an inference, denoted by o_maxvar. By default, this number is equal to twice the number of variables that appear in any base clause. This heuristic sacrifices completeness. However, the user can set the value of o_maxvar before beginning the search for a proof to any value between 0 and 32.

## 9.9 Ordering Clauses at Each Node

As clauses are generated at each node and before they are searched, THEO, like HERBY, (see Section 7.2.2.1) orders them from shortest to longest. This is done because short clauses are more likely than long ones to yield a contradiction (knowing nothing else about the clauses). This heuristic does not sacrifice completeness.

## 9.10 A Hash Table that Stores Information About Clauses

THEO uses a large hash table called **clause_hash_table** to store information about clauses generated during the search. This information is used to make inferences. The array has a default size of 2 Meg entries ($2^{21}$ entries), with each entry occupying four words. The size of the array can be changed by the user but must be set to a power of two. The first two words of an entry are a 64-bit **clause hash code**, and the next word records the number of literals in the clause and the level in the tree at which the clause was found. The fourth word records the iteration on which the clause was found.

### 9.10.1 Assigning a hash code to a literal

A literal L consists of a string of symbols beginning with or without a negation symbol, followed by a symbol representing a predicate, followed in turn by symbols representing the terms (i.e., the functions, constants, and variables in the literal). Assume the literal has k terms, and that it appears as $PT_1T_2 \ldots T_k$ (or $\sim PT_1T_2 \ldots T_k$). A **hash function** takes this description of a literal — the negation symbol if it is present — followed by the predicate symbol, followed by k symbols representing the terms, and maps them into a 64-bit number called the **literal hash code** and denoted HC(L).

In theory, a hash function is supposed to produce different hash codes for different strings of symbols. In practice, there may be more than $2^{64}$ different strings, making this goal impossible. **Hash errors** are said to occur when two different literals are assigned the same 64-bit hash code. A good hash function makes the probability of this happening very small, as seems to be the case with the function used by THEO.

THEO assigns a hash code to a literal with k terms by assigning a random 64-bit number to the predicate and to each term $T_i$ based on its position in the string. Then, the exclusive-or of the random numbers assigned to the predicate name and to the k terms is computed. Finally, if the literal begins with a negation symbol, the result of this exclusive-or operation is negated. The hash codes of two complementary literals are thus equal in magnitude but opposite in sign, and their sum is zero.

Predicate names are assigned numerical values based on the order in which they first appear in a theorem. The numerical value assigned to a function name is determined in a two-step procedure. In the first step, as the theorem is read from disk, function names are assigned integers beginning at one and in the order that they first appear in the theorem. In the second step, the names are sorted so that the constants appear at the top of the list, followed by the other functions. For example, for S31WOS7.THM in Figure 9.9, the five functions e, g, f, a, b are initially assigned values 1, 2, 3, 4, 5, respectively. In the second step, they are sorted and renumbered 1, 4, 5, 2, 3, respectively.

The 64-bit random numbers are found in an array called rand_array. This array is created when THEO reads in the theorem. It has 200 columns and N rows, where N is equal to F + G + MAXVARS, where F is the number of functions in the theorem, G is the number of predicates, and MAXVARS is 32. N is set to this value because it must be equal to the maximum number of different symbols that can appear in the clause. The first F + G rows of rand_array correspond to the F function and G predicate names in the theorem; the remaining MAXVARS rows correspond to the different variable names that might be assigned to the variables that appear in a clause.

```
Axioms
1  P(x,f(a)) | ~Q(y,f(x))
2  ~P(x,y) | ~Q(z,y)

Negated conclusion
3  Q(a,x)
```

Figure 9.8. HASH.THM.

For example, consider HASH.THM in Figure 9.8. There are two functions, a and f, so F = 2. There are two predicates P and Q, so G = 2. Thus, rand_array is of size (2 + 2 + 32)*200 = 7200. Row 0 corresponds to a, row 1 to f, row 2 to P, row 3 to Q, and the remaining 32 rows to each of the possible 32 variables beginning with x, y, z, etc. Before beginning to search for a proof, THEO generates 7200 64-bit (two words) random numbers for the 7200 elements of rand_array.

The hash code for a literal L, denoted HC(L), is then given by:

$$HC(L) = SIGN * C(P) \oplus \{ \bigoplus_{1 \le i \le k} rand\_array(T_i, i) \}$$

where SIGN is +1 if the literal is not negated and –1 if it is, C(P) is the random number assigned to predicate P, $T_i$ is the i-th of k terms in the string of terms, ⊕ is the exclusive-or function. The random number assigned to predicate Q — that is, C(Q) — in HASH.THM is rand_array(3,0). The second literal of the clause 1 P(x,f(a)) | ~Q(y,f(x)) in HASH.THM is assigned a hash code −1*[rand_array(3,0) ⊕ rand_array(5,1) ⊕ rand_array(1,2) ⊕ rand_array(4,3)].

Assigning hash codes to literals makes determining whether two literals are identical an easy task — just compare their hash codes. Determining whether they are complementary, but otherwise identical, is also easy: just see if their hash codes add to zero.

### 9.10.2 Assigning a hash code to a clause

The hash code of a clause is calculated once the hash code of each literal has been determined. The hash codes of the literals are ordered from largest to smallest (ensuring P(a) | P(b) has the same hash code as P(b) | P(a)) and then multiplied together by a complex function to form the clause hash code.

Just as assigning hash codes to literals makes determining whether two literals are identical an easy task, so does assigning hash codes to clauses. Two clauses are identical if their hash codes are identical. Two clauses, logically the same such as P(x) | Q(y) and P(x) | Q(z), however, are assigned different hash codes and considered different clauses by THEO. We will have more to say on this problem later.

### 9.10.3 Entering clauses in clause_hash_table

When a clause C is generated at some node, information about the clause is stored in an entry in clause_hash_table. The clause hash code of C tells THEO where to store the information: THEO uses the least significant B bits of the clause hash code as the address at which to store the information. B is equal to the log (base 2) of the number of entries in clause_hash_table.

THEO generates several thousand clause hash codes per second, and the two million entry hash table can fill up after several minutes. As the hash table fills up, it becomes increasingly more difficult to find an empty location. If THEO attempts to store information about some clause at a location where information on another clause has already been stored, THEO will **probe** as many as 11 other locations. If THEO is still unable to find an empty location, it replaces the clause (of the 12 probed) that has the maximum number of literals. A **hash table overflow** is said to occur. This usually only happens when the hash table is approximately 90% full. For theorems that take a long time to prove, the hash table gradually decreases in effectiveness during the search for a proof.

## 9.11 Using Entries in clause_hash_table

THEO avoids searching certain clauses that normally would be searched in the extended search region when the extended search strategy is used. Specifically, on the nth iteration, if a clause C is generated at level n or deeper, and if C was previously generated at level n or less, THEO does not search successors of C because it statistically does not seem to pay off and because it does not interfere with the completeness of the search strategy. clause_hash_table contains the information on C used to make this decision.

## 9.12 Unit Clauses

Unit clauses are hashed into clause_hash_table just like any other clause. They are used, however, in two special ways to speed up the search for a proof, as described in Sections 9.12.1 and 9.12.2.

### 9.12.1 Obtaining a contradiction

When a search for a proof begins, THEO enters Phase 1. Whenever a unit clause is found during this phase, say clause U, with hash code HC(U), clause_hash_table is searched to see whether the hash code of the complement of U — that is, –HC(U) — is an entry in the table. If so, THEO realizes that (1) a proof exists and that part of it, the **main line**, has been found, and that (2) THEO must repeat the search to find the path that led to the clause with the hash code –HC(U). THEO enters Phase 2 to do this. In Phase 2, the search carried out in Phase 1 is repeated, looking for the unit clause with hash code –HC(U) and the path to that clause. When the path is found, the complete proof can be assembled.

This use of clause_hash_table allows THEO to find proofs that are more complex than linear. It allows THEO to find proofs consisting of two paths: on the nth iteration, THEO can find a proof whose length is as great as 2n. The extended search strategy allows even longer proofs to be found. This is illustrated by the proof of S31WOS7.THM found by an early version of THEO, one that did not incorporate the ideas of the coming several sections. The theorem is presented in Figure 9.9. THEO found the literal ~p(g(a),e,b) early in the first iteration and stored it in clause_hash_table. Note that extended search allowed this path of length three to be found on the first iteration. Later, the path to p(g(a),e,b) was searched, and when this unit clause was arrived at, it was found that the hash code of its complement, –HC(p(g(a),e,b),

```
Axioms
 1 p(e,x,x)
 2 p(g(x),x,e)
 3 p(x,y,f(x,y))
 4 ~p(x,y,u) | ~p(y,z,v) | ~p(u,z,w) | p(x,v,w)
 5 ~p(x,y,u) | ~p(y,z,v) | ~p(x,v,w) | p(u,z,w)
 6 Equal(x,x)
 7 ~Equal(x,y) | Equal(y,x)
 8 ~Equal(x,y) | ~Equal(y,z) | Equal(x,z)
 9 ~p(x,y,u) | ~p(x,y,v) | Equal(u,v)
10 ~Equal(u,v) | ~p(x,y,u) | p(x,y,v)
11 ~Equal(u,v) | ~p(x,u,y) | p(x,v,y)
12 ~Equal(u,v) | ~p(u,x,y) | p(v,x,y)
13 ~Equal(u,v) | Equal(f(x,u),f(x,v))
14 ~Equal(u,v) | Equal(f(u,y),f(v,y))
15 ~Equal(u,v) | Equal(g(u),g(v))
16 p(x,e,x)
17 p(x,g(x),e)
18 p(a,b,e)

Negated conclusion
19 ~p(b,a,e)
```

Figure 9.9. S31WOS7.THM.

was in clause_hash_table. Again, note that extended search allowed this path of length four to be found on the first iteration. THEO knew a proof existed at this point:

```
20: (18a,4b)  ~p(x,a,y) | ~p(y,b,z) | p(x,e,z)
  21: (20a,2a)  ~p(e,b,x) | p(g(a),e,x)
    22: (21a,1a)  p(g(a),e,b)
      23: (22a,HC(~p(g(a),e,b)) Ø
```

The notation of the last line indicates that a contradiction was found when the unit clause p(g(a),e,b) was found to have the hash code of its complement, which was previously entered in clause_ hash_ table. A repeat of the search in Phase 2 was necessary to find the sequence of inferences leading to ~p(g(a),e,b). This is illustrated in Figure 9.10; the dashed line leading to the literal ~p(g(a),e,b) indicates that the path to ~p(g(a),e,b) was not known when the path to p(g(a),e,b) was found. Phase 2, during which the search was exactly repeated, found this path:

```
20: (19a,5d)  ~p(x,y,b) | ~p(y,a,z) | ~p(x,z,e)
  21: (20c,2a) ~p(g(x),y,b) | ~p(y,a,x)
    22: (21b,1a)  ~p(g(a),e,b)
```

After renumbering clauses from the main line, the complete proof was assembled together:

```
20: (19a,5d)  ~p(x,y,b) | ~p(y,a,z) | ~p(x,z,e)
  21: (20c,2a) ~p(g(x),y,b) | ~p(y,a,x)
    22: (21b,1a)  ~p(g(a),e,b)
23: (18a,4b)  ~p(x,a,y) | ~p(y,z,b) | p(x,e,z)
  24: (23a,2a)  ~p(e,b,x) | p(g(a),e,x)
    25: (24a,1a)  p(g(a),e,b)
      26: (25a,22a)  Ø
```

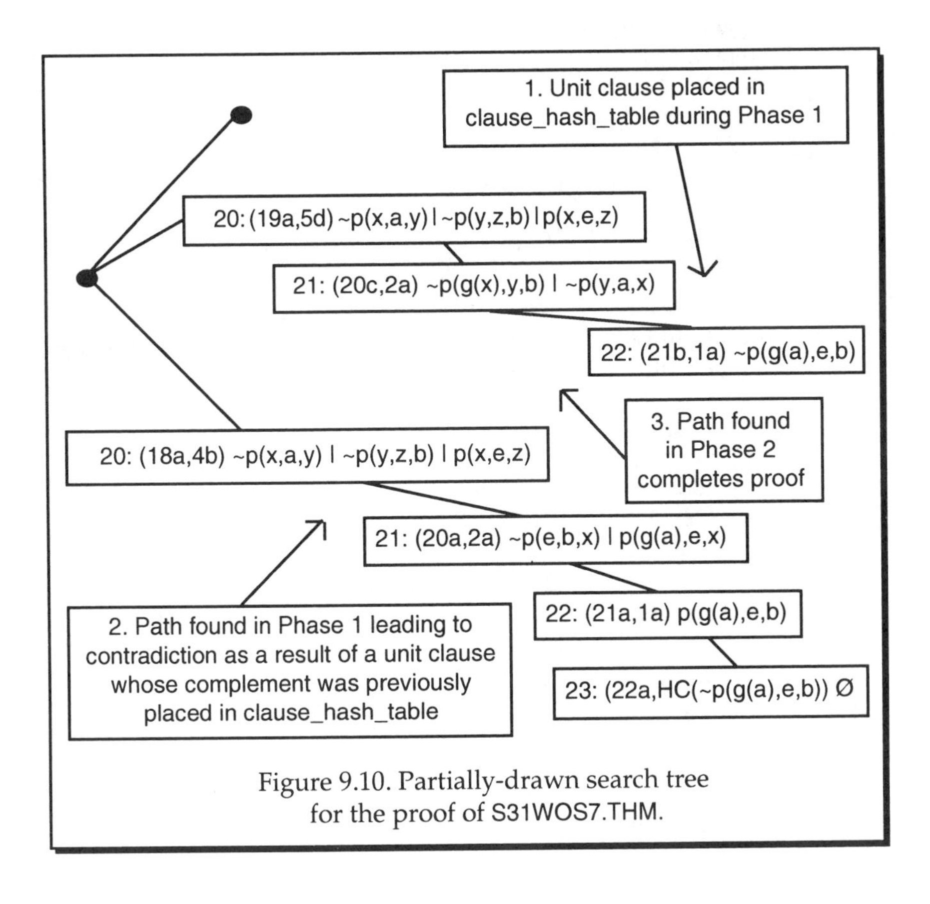


Figure 9.10. Partially-drawn search tree for the proof of S31WOS7.THM.

## 9.12.2 Unit Hash Table Resolutions

Inferences in THEO are actually formed in two stages. In the first stage, a binary resolvent of two clauses is formed or a binary factor of a single clause is formed, as Chapter 4 describes. In the second stage, the hash code of each literal of the new inference is compared with entries in clause_hash_table. For each match found with the negative hash code of a previously entered unit clause, that literal can be deleted from the clause in what is called a **unit hash table resolution**. Several literals may be so deleted. If a proof is found later, it is necessary to repeat the search to find the sequence of inferences that led to each of the hash codes used in the unit hash table resolutions.

S39WOS15.THM, in Figure 9.11, illustrates unit hash table resolutions. As with S31WOS7.THM in the previous section, an early version of THEO was used. One unit hash table resolution occurred in the proof. THEO knew it had a proof when it found the path leading to the unit clause p(a,g(g(b)),c):

```
24: (22a,5a)  ~p(b,x,y) | ~p(a,y,z) | p(c,x,z)
 25: (24a,17a) ~p(a,e,x) | p(c,g(b),x)
  26: (25a,16a) p(c,g(b),a)
   27: (26a,5a)  ~p(g(b),x,y) | ~p(c,y,z) | p(a,x,z)
    28: (27a,17a) ~p(c,e,x) | p(a,g(g(b),x)
     29: (28a,16a) p(a,g(g(b)),c)
```

whose complement had been found previously and placed in clause_hash_table. THEO was in Phase 1 until this time. It then entered Phase 2 where it knew a proof existed. In Phase 2, the search may have to be repeated several times to construct the proof. Here, the second pass found the path to ~p(a,g(g(b)),c):

```
24: (21a,18c) ~p(x,g(b),y | ~o(x) | o(y)
 25: (24a,1a) o(g(b)) | <~oe>
  26: (25a,18c) ~p(x,g(g(b)),y) | ~o(x) | o(y)
   27: (26c,23a) ~p(x,g(g(b)),c) | ~o(x)
    28: (27b,20a) ~p(a,g(g(b)),c)
```

The second inference on this path says that clause 25 was formed by resolving the first literal of clause 24 with the first literal of clause 1 and then eliminating literal ~o(e) as a result of a unit hash table resolution. A third search was necessary to find the path to o(e). THEO found:

```
24: (21a,18c) ~p(x,g(b),y) | ~o(x) | o(y)
 25: (24a,17a) ~o(b) | o(e)
  26: (25a,21a) o(e)
```

These three paths were assembled together to yield the proof of length sixteen shown in Figure 9.11. The use of clause_hash_table in this second way gives THEO the power to find very long proofs.

```
Axioms
 1 p(e,x,x)
 2 p(g(x),x,e)
 3 p(x,y,f(x,y))
 4 ~p(x,y,u) | ~p(y,z,v) | ~p(u,z,w) | p(x,v,w)
 5 ~p(x,y,u) | ~p(y,z,v) | ~p(x,v,w) | p(u,z,w)
 6 Equal(x,x)
 7 ~Equal(x,y) | Equal(y,x)
 8 ~Equal(x,y) | ~Equal(y,z) | Equal(x,z)
 9 ~p(x,y,u) | ~p(x,y,v) | Equal(u,v)
10 ~Equal(u,v) | ~p(x,y,u) | p(x,y,v)
11 ~Equal(u,v) | ~p(x,u,y) | p(x,v,y)
12 ~Equal(u,v) | ~p(u,x,y) | p(v,x,y)
13 ~Equal(u,v) | Equal(f(x,u),f(x,v))
14 ~Equal(u,v) | Equal(f(u,y),f(v,y))
15 ~Equal(u,v) | Equal(g(u),g(v))
16 p(x,e,x)
17 p(x,g(x),e)
18 ~o(x) | ~o(y) | ~p(x,g(y),z) | o(z)
19 ~o(x) | ~Equal(x,y) | o(y)
20 o(a)
21 o(b)
22 p(a,b,c)

Negated conclusion
23 ~o(c)

Proof:
24: (21a,18c) ~p(x,g(b),y) ~ | o(x) |  o(y)
  25: (24a,17a) ~o(b) | o(e)
    26: (25a,21a) o(e)
27: (21a,18c) ~p(x,g(b),y)  | ~o(x) | o(y)
  28: (27a,1a)  ~o(e) | o(g(b))
    29: (28a,26a) o(g(b))
      30: (29a,18c) ~p(x,g(g(b)),y)  | ~o(x) | o(y)
        31: (30c,23a) ~p(x,g(g(b)),c) | ~o(x)
          32: (31b,20a) ~p(a,g(g(b)),c)
33: (22a,5a)  ~p(b,x,y) | ~p(a,y,z) | p(c,x,z)
  34: (33a,17a) ~p(a,e,x) | p(c,g(b),x)
    35: (34a,16a) p(c,g(b),a)
      36: (35a,5a)  ~p(g(b),x,y) | ~p(c,y,z) | p(a,x,z)
        37: (36a,17a) ~p(c,e,x) | p(a,g(g(b)),x)
          38: (37a,16a) p(a,g(g(b)),c)
            39: (38a,32a) Ø
```

Figure 9.11. A proof of S39WOS15.THM illustrating a unit hash table resolution.

## 9.13 Assigning Hash Codes to Instances and Variants of Unit Clauses

In Section 9.12, two uses of clause_hash_table were described. In the first case, a proof is obtained if two identical complementary unit clauses are found in the search tree. The opposite-signed values of their hash codes signal this. In the second case, a unit hash table resolution can eliminate a literal in a clause if the identical complementary unit clause was previously found.

However, as described thus far, THEO would not detect that it had a proof if it had the hash code for p(x) in clause_hash_ table and then found the clause ~p(a) in the search tree. It would also not realize that if it had placed the hash code for p(x) in clause_hash_ table and then later found ~p(y) that it had a proof. It also would not realize that ~p(y) could be eliminated in the clause q(x) I ~p(y) through a unit hash table resolution. Lastly, it would not realize that if it had earlier found p(x) in the search, it should not extend the search on the clause p(a) I q(x).

To permit THEO to handle situations like these, code was added that calculates more than one hash code for each unit clause and stores these additional hash codes and related information in clause_hash_table. In other words, when a unit clause is generated, THEO generates a number of hash codes of different instances and variants — the same clause, except the variables are assigned different names — and stores these hash codes in clause_hash_ table as the following five steps describe:

(1) The hash code of the unit clause is generated and it, along with information about the clause, is stored in clause_hash_table. This has already been discussed in Section 9.10.

Recall from Section 4.5 that the variables of a clause are said to be in normal order if when they first appear are ordered x,y,z, ... . For example, the variables of p(x,y,f(y,x)) are in normal order as are those of p(x,g(x),y); on the other hand, the variables of r(x,z), r(y,x), and r(y,y) are not in normal order. When THEO forms a resolvent that has more than one literal, the first literal of the resolvent is always in normal form. The variables of the other literals may or may not be. Because a unit hash table resolution can cause a literal in an inference to be resolved away — chopped off essentially — the variables of a unit clause thus created may not be in normal order. For example, suppose THEO generated the resolvent P(x,y) as some point in the search and then later generated the resolvent ~P(x,y) I Q(y,x). The literal ~P(x,y) would be resolved away as a result of a unit hash table resolution, leaving the unit clause Q(y,x), with variables not in normal order.

(2) The variables of the unit clause are converted to normal order if the clause has at least one variable and the first literal in the clause has been resolved away by a unit hash table resolution. Thus, Q(y,x) would be converted to Q(x,y) in the previous example. The hash code for this mdified unit clause is calculated and then entered in clause_hash_table.

(3) If there is exactly one variable in the unit clause and k constants in the theorem, the k hash codes of the k instances of the unit clause, which have that variable replaced by each constant in the theorem, are calculated and entered in clause_hash_table. The hash code of the unit clause with the variable x replaced by y is also calculated and entered in clause_hash_table.

(4) If there are exactly two variables in the unit clause and k constants in the theorem, the 2*k hash codes of the 2*k instances of the unit clause in which one of the two variables has been replaced by one of the k constants in the theorem are calculated and entered in clause_hash_table.

(5) If there are two or more variables, say m, in the unit clause and k constants in the theorem, the k*k*m*(m – 1)/2 hash codes of the k*k*m*(m – 1)/2 instances of the unit clause in which for all combinations of the m variables, all k*k combinations of the constants in the theorem are calculated and entered into clause_hash_table.

Thus, for a unit clause with four variables and for a theorem with five constants, a total of 1 + 1 + 0 + 0 + 5*5*4*3/2 = 152 hash codes would be calculated. However, the number of variables in unit clauses is usually small, and such a large number of hash codes is typically not generated.

For example, there would be 1 + 0 + 0 + 2*4 + 4*4*2*1/2 = 25 hash codes generated for the unit clause 3 p(x,y,f(x,y)) in S39WOS15.THM. The hash codes would correspond to the following instances: (1) p(x,y,f(x,y)), (2) p(a,x,f(a,x)), (3) p(b,x,f(b,x)), (4) p(c,x,f(c,x)), (5) p(e,x,f(e,x)), (6) p(x,a,f(x,a)), ... (10) p(a,a,f(a,a)), (11) p(a,b,f(a,b)), ... (25) p(e,e,f(e,e)).

## 9.14 Other Hash Codes Generated

During the course of the search, if THEO generates a clause with two or three literals and one variable, such as, for example, P(x) | Q(f(x)) | R(h(x)), it searches the hash table to see if the negative hash codes of ground instances of two of the literals of the clause have been previously entered into clause_hash_table where the variable x was replaced by the same constant.

All constants in the theorem are checked. If so, THEO then enters in clause_hash_table the hash code of the third literal with its variable replaced by the same constant. For example, if THEO generated P(x) | Q(f(x)) | R(h(x)) and if THEO had previously entered hash codes for ~P(a) and ~Q(f(a)) in clause_hash_table, THEO would then enter the hash code for R(h(a)) in clause_hash_table. THEO actually, after finding the hash codes for the first two literals, first checks clause_hash_table to see if the negative hash code of the third literal, ~R(h(a)) in our example, was previously entered. If so, THEO realizes it has found that a proof exists. If not, THEO then enters the hash code of this instantiated literal, R(h(a)) in our example, in clause_hash_table.

## 9.15 The Use of S-Subsumption to Restrict the Search Space

Determining whether one clause subsumes another is a computationally expensive procedure. Although computationally expensive, it is particularly effective in reducing the search space to be examined when hunting for the NULL clause. THEO takes a middle ground approach—avoiding a test for general subsumption — by testing instead for s-subsumption. It is much easier to test for this type of subsumption; all that must be done is to compare the hash codes of the literals.

At each node and as each clause is generated, it is tested to see whether it is s-subsumed by some other clause generated at that node, by a base clause or by a clause on the continuation leading to the node. If so, the clause is eliminated. This heuristic does not affect completeness.

## 9.16 Extending the Search on Merge Clauses

Search is extended one level deeper on all lines that have a merge clause at the search horizon. This encourages THEO to explore lines with merge clauses. This heuristic does not affect completeness.

## 9.17 Extending the Search on Clauses from the Negated Conclusion

Search is extended one level deeper on all lines that are descendants of a clause from the negated conclusion. However, THEO will not increment thedepth of search by two as a result of having a merge clause at the end of a path and a clause from the negated conclusion somewhere on the path. This heuristic does not affect completeness.

## 9.18 Do Not Factor Within the Search Horizon

Except in the extended search region, form factors only of clauses that are descendants of the negated conclusion. It is left for the reader to determine whether this heuristic affects completeness.

## 9.19 Clauses in the Extended Search Region Must Have a Constant

If a clause in the extended search region has no constants and the theorem has one or more constants, eliminate the clause. Since this is done only beyond the search horizon, it does not affect completeness.

## 9.20 In the Extended Search Region, Do Not Search the Single Inference of a Long Clause

In the extended search region, if a clause with three or more literals has only one inference, terminate the search at that node. The single inference is highly unlikely to lead to a unit clause or a contradiction. Because this is done only beyond the search horizon, it does not affect completeness.

## 9.21 Eliminate Tautologies

If a tautology is generated, it is deleted. This heuristic does not affect completeness.

## 9.22 Special Consideration of the Predicate Equal

THEO has two heuristics that deal explicitly with the Equal predicate. (1) If a resolvent has a literal Equal(term1,term1), the resolvent is considered TRUE and dropped. (2) If a resolvent has a literal ~Equal(term1,term1), the literal is dropped from the clause just as is done when a unit hash table resolution is performed. This heuristic does not affect completeness.

## 9.23 Assembling the Proof

The array named cl is used to store information about the inferences in the lines of a proof as they are found. Each row of the array stores the number of the parent clauses and the indices of the resolved literals. When a hash code from the hash table was used in a resolution, the hash code is also stored. An example of how this information is stored and used to assemble the proof is presented in Section 10.7.

## 9.24 eliminate: Simplifying the Base Clauses

The base clauses of a theorem can often be simplified before beginning a search for a proof. This is done by THEO in Phase 0 in a function called eliminate. A set of clauses can be thought of as simplified if the new modified set makes the search for the NULL clause proceed more quickly. The four functions that are called by eliminate (in elim.c) to simplify a set of clauses generally do this. Two functions eliminate clauses that cannot help in finding a contradiction, a third replaces s-subsumed clauses with resolvents of these clauses, while a fourth eliminates "long" clauses.

The function eliminate loops through Steps 0–3 (see later.) until no base clauses are changed in Steps 1–3. If the set of base clauses is modified in any of the Steps 1–3, eliminate jumps directly back to Step 0. When no modifications occur, eliminate goes on to execute Step 4 and then terminates. THEO then enters Phase 1, where the search for a proof is carried out using the modified set of base clauses. Let BASE denote the number of base clauses.

Step 0: Generate all resolvents of the base clauses.

Step 1: (Uncomplemented literal) If some base clause C has a literal that cannot be resolved with any other literal in the base clauses, delete C, set BASE = BASE – 1, and return to Step 0. (C cannot contribute to the proof.)

Step 2: (Eliminate one base clause s-subsumed by another) If some base clause, say B1, is s-subsumed by some other base clause, say B2, delete B1, set BASE = BASE – 1, and then return to Step 0. (If a proof exists using B1, a simpler one exists using B2.)

Step 3: (If a resolvent of a base clause s-subsumes its parent, eliminate the parent) If some clause C1 generated in Step 0 s-subsumes one of its parents B1, replace B1 with C1 and return to Step 0. (If a proof exists using clause B1, a simpler one exists using C1.)

Step 4: (Eliminate clauses that have too many literals) Delete all clauses that have more than o_maxlit + 1.

## 9.25 Summary of THEO's Strategy

THEO passes through four phases on the way to obtaining a proof. In summary, they are:

Phase 0: Attempt to simplify the clauses of a theorem with the function eliminate.

Phase 1: Carry out a sequence of iteratively deepening searches. Each search is carried out in a depth-first manner looking for a linear proof. At each node in the search tree, inferences are generated as described by the pseudo C code in Figure 9.2. Some of the inferences so generated are eliminated or filtered out when looking for a linear-merge proof (normally ON),or when looking for a linear-nc proof (normally OFF). The extended search strategy (normally ON) encourages THEO to follow promising lines more deeply in the tree. Some clauses are eliminated because they have too many literals, terms, or variables. Inferences that are tautologies are eliminated. Inferences are ordered from shortest to longest for search at each node as they are generated. Inferences are eliminated if their hash codes are found in clause_hash_table and they meet certain conditions. They are also eliminated if they are found to be s-subsumed by certain other clauses. By using clause_hash_table,THEO has the ability to find proofs more complex than linear. Several additional heuristics focus the search. In particular, the search is extended one additional level if the inference at the search horizon is a merge clause. It is also extended one level if the inference at the search horizon is a descendant of the negated conclusion. Most factors are not generated within the search horizon, while clauses beyond the search horizon must contain at least one constant (if the theorem has one or more constants). Also, beyond the search horizon, do not search the single inference of a long clause. Lastly, some special consideration is given to the Equal predicate. The processing of an inference is illustrated in Figure 9.12.

Phase 2: Repeat the search carried out in Phase 1 as many times as necessary to find and assemble all the lines of the proof.

Phase 3: Check the proof to be sure no hash error happened, and then print out the results.

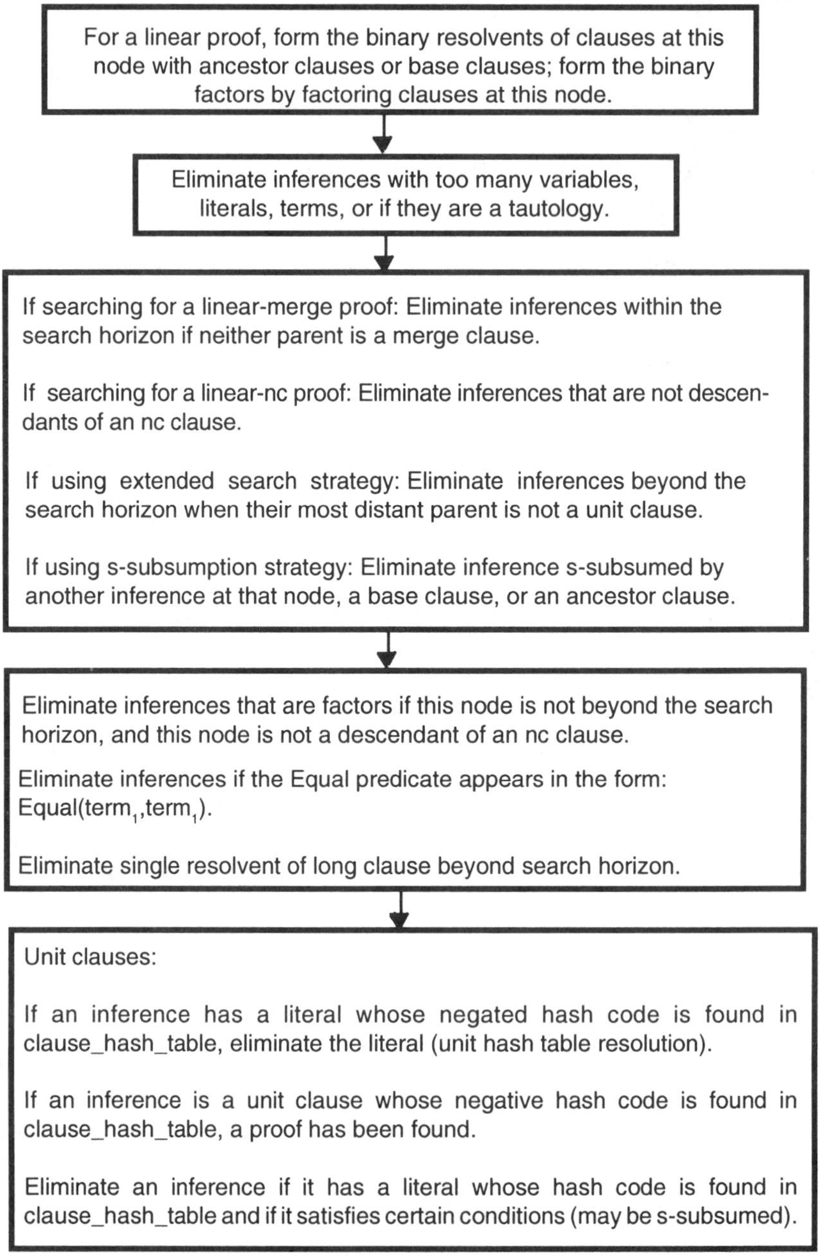

For a linear proof, form the binary resolvents of clauses at this node with ancestor clauses or base clauses; form the binary factors by factoring clauses at this node.

↓

Eliminate inferences with too many variables, literals, terms, or if they are a tautology.

↓

If searching for a linear-merge proof: Eliminate inferences within the search horizon if neither parent is a merge clause.

If searching for a linear-nc proof: Eliminate inferences that are not descendants of an nc clause.

If using extended search strategy: Eliminate inferences beyond the search horizon when their most distant parent is not a unit clause.

If using s-subsumption strategy: Eliminate inference s-subsumed by another inference at that node, a base clause, or an ancestor clause.

↓

Eliminate inferences that are factors if this node is not beyond the search horizon, and this node is not a descendant of an nc clause.

Eliminate inferences if the Equal predicate appears in the form: Equal($term_1$,$term_1$).

Eliminate single resolvent of long clause beyond search horizon.

↓

Unit clauses:

If an inference has a literal whose negated hash code is found in clause_hash_table, eliminate the literal (unit hash table resolution).

If an inference is a unit clause whose negative hash code is found in clause_hash_table, a proof has been found.

Eliminate an inference if it has a literal whose hash code is found in clause_hash_table and if it satisfies certain conditions (may be s-subsumed).

Figure 9.12. Rules for generating inferences at a node.

# Exercises for Chapter 9

9.1. Consider B.THM in Figure 9.13. Similar to Figure 9.3 and using the pseudocode in Figure 9.2, draw the search trees generated by THEO when searching for a linear proof. How many iterations are carried out? How many clauses are generated?

9.2. Consider B.THM and C.THM. Similar to Figure 9.6, draw the search trees generated by THEO when searching for a linear-merge-nc proof of each theorem. For each theorem, how many iterations are carried out and how many clauses are generated?

Axioms
1 P(x) | ~Q(x)
2 Q(a) | P(a)

Negated conclusion
3 ~P(x)

Figure 9.13a. B.THM.

Axioms
1 ~P(x) | Q(a)
2 ~Q(a) | ~R(x)
3 R(a) | S(a)
4 ~R(x) | P(y)

Negated conclusion
5 ~S(x)

Figure 9.13b. C.THM.

9.3. Again, consider B.THM and C.THM, and similar to Figure 9.7, draw the search trees generated by THEO when searching for a linear-merge-nc proof using the extended search strategy for each theorem. For each theorem, how many iterations are carried out and how many clauses are generated?

9.4. Consider HASH.THM in Figure 9.8. Suppose a naive random number generator assigned the entries in the 36*200 element rand_array as rand_array[i][j] = 36 j + i. What is the value of the hash code that THEO would assign to each of the five literals of the base clauses?

9.5. Prove the following statement or show a counterexample: If a theorem has no more than X literals in any base clause, then a proof exists in which no inference has more than X – 1 literals.

9.6. Consider S39WOS15.THM in Figure 9.11. Specify the number of hash codes entered in clause_hash_table for each of the ten unit base clauses.

THEO

# 10 Using THEO

This chapter describes how to use THEO to prove theorems. The input to THEO is a text file formatted either in its own HERBY/THEO format or in the format of the TPTP Problem Library. This is described in Section 10.1. THEO saves the results in an output file, as described in Section 10.2. Options that the user can control are described in Section 10.3. User interaction during the search is described in Section 10.4. The printout produced by THEO when proving a theorem is presented in Section 10.5. The printout when using the d1 option is presented in Section 10.6. This option allows the user to follow THEO's progress in constructing a proof. Proving a set of theorems without user intervention is described in Section 10.7.

## 10.1 Proving Theorems with THEO: The Input File

A theorem that THEO is asked to prove must be in a text file. It must be formatted in either HERBY/THEO format or in the format used by the TPTP Problem Library.

In HERBY/THEO format, the character string negated_conclusion must be on a line separating the given axioms from the negated conclusion. Each clause must be on a single line. Comments are considered to be any characters on a line following a semicolon. The various theorems provided with this package should be examined for more specifics.

For theorems in TPTP Problem Library format, there may be axioms in include files. These files must be in a subdirectory of the directory containing the theorem or located two levels higher. Again, the theorems in the TPTP Problem Library should be examined for specifics.

To prove the theorem called S39WOS15.THM in HERBY/THEO format in the directory called theorems using default options, simply type:

theo

THEO responds with:

Enter the name of the theorem (Type ? for help):

You should then enter the name of the file that contains the theorem:

theorems/S39WOS15.THM

THEO will then attempt to find a proof. After one hour or after a time bound established by you has been exceeded and no proof found or after thirty iterations have been completed, THEO will give up. During the course of attempting to find a proof and after the search terminates for one reason or another, information is also printed out to disk.

## 10.2 THEO's Convention on Naming the Output File

The output file is assigned the same name as the input file, except a ".t" or a ".Tx" suffix replaces the first period (".") and anything following the first period in the name of the input file or is appended to the end of the name of the input file.If THEO does not find a proof, a .t suffix is added. If a proof is found, a .Tx suffix is added, where x is the time in seconds required to find a solution. For example, if the input file is S39WOS15.THM and a proof is found in 46 seconds, the output file is named S39WOS15.T46. If no proof was found, the results of the search would have been placed in S39WOS15.t. This makes it easy to determine whether a proof to a theorem has been found: you can simply look at the name of the file produced.

## 10.3 The Options Available to the User

There are 18 options available to the user when working with THEO. They must be selected when entering the name of the theorem. The options fall into three broad categories: (1) options that determine the search strategy used by THEO, (2) options that determine the information displayed during and after the search, and (3) an option that can be used to prove a set of theorems. The default value for each option is shown underlined in what follows. The user can override this value by entering the abbreviation for the option followed by its preferred value. Options must follow the name of the theorem. A space must separate each option.

## 10.3.1 Options that determine the search strategy

OPTION 1: Perform the simplification phase (Phase 0) (see Section 9.24).

| | |
|---|---|
| k1 | Perform the simplification phase before searching for a proof. |
| k0 | Do not perfom the simplification phase. |
| k2 | Perfom only the simplification phase. |

OPTION 2: Search for a linear-merge proof (see Section 9.2).

| | |
|---|---|
| m1 | Search for a linear-merge proof. |
| m0 | Do not search for a linear-merge proof. |

OPTION 3: Search for a linear-nc proof (see Section 9.3).

| | |
|---|---|
| z0 | Do not search for a linear-nc proof. |
| z1 | Search for a linear-nc proof. |

OPTION 4: Use clause_hash_table (see Sections 9.10–8.14).

| | |
|---|---|
| h1 | Use clause_hash_table. |
| h0 | Do not use clause_hash_table. |

OPTION 5: Use the extended-search strategy (see Section 9.5).

| | |
|---|---|
| n1 | Use the extended-search strategy during the search. |
| n0 | Do not use the extended-search strategy during the search. |

OPTION 6: Use simple subsumption (see Section 9.15).

| | |
|---|---|
| b1 | Use simple subsumption to eliminate clauses. |
| b0 | Do not use simple subsumption to eliminate clauses. |

OPTION 7: Bound the number of literals in a clause (see Section 9.6).

| | |
|---|---|
| li | Eliminate clauses with more than i literals during the search for a proof. Default value is determined as described in Section 9.6. |

OPTION 8: Bound the number of terms in a literal (see Section 9.7).

| | |
|---|---|
| xi | Eliminate clauses with a literal having more than i terms. Default value is determined as described in Section 9.7. |

OPTION 9: Bound the number of distinct variables in a clause (see Section 9.8).

| | |
|---|---|
| vi | Eliminate clauses with more than i variables. Default value is determined as described in Section 9.8. |

OPTION 10: Set the size of clause_hash_table.

| | |
|---|---|
| axx...x | Set the size of clause_hash_table to xx...x where xx...x is a power of two. The default value of xx...x is 2097152 (2 Meg entries). |

OPTION 11: Select TPTP Problem Library format for the input clauses.

tptp   If tptp is included, THEO assumes the theorem is in TPTP Problem Library format.

OPTION 12: Set the maximum time given to THEO to find a proof.

ti   Search for a proof for at most i seconds.
The default value is 3600 seconds, or one hour.

Option 1, normally ON, usually has only positive effects. Option 2, restricting the search to a linear-merge proof, normally ON, narrows the search but often results in longer proofs than when it is not used. It does not sacrifice completeness. Option 3, restricting search to a linear-nc proof, also has been shown not to sacrifice completeness, but THEO seems more effective when the strategy is not used and thus this strategy is normally OFF. Again, using this strategy sometimes results in longer proofs than when not using it. Option 4, the use of clause_hash_table, normally ON, permits finding proofs having more than one path; it does not affect completeness. Option 5, the extended-search strategy, normally ON, allows finding proofs longer than n inferences on the nth iteration; it does not sacrifice completeness. Option 6, using simple-subsumption, normally ON, eliminates some unnecessary clauses and does not affect completeness. Option 7, normally ON, limits the number of literals in each inference. Option 8, normally determined by THEO, limits the number of terms in a predicate, sacrificing completeness. Option 9, normally determined by THEO, bounds the number of variables in a clause, sacrificing completeness. Option 10, placing a time limit on the search, certainly sacrifices completeness. Although the default values used by THEO for Options 7–9 can be overridden by the user, data structures place absolute upper bounds on these parameters. Option 10 can be used to change the size of clause_hash_table. Option 11 is used when the theorem is in the TPTP Problem Library.

### 10.3.2 Options that determine the information observed by the user

The user can observe the steps carried out during the simplification phase and during the search phases by using the following options.

OPTION 13: Print the steps performed in Phase 0.

<u>e0</u>   Do not print the steps performed in the simplification phase.

e1   Print the steps carried out.

OPTION 14: Print the tree as the search progresses in Phases 1 and 2.

pi Print the first i levels of the tree as it is searched. Default value is p0.

OPTION 15: Step from node to node in the search tree in Phases 1 and 2.

sj Stop and wait for instructions after printing each clause at the first j levels of the tree. This option is normally used in conjunction with the Option 14. For example, p5 s2 instructs THEO to print the first five levels of the tree while stopping after printing each clause at the first two levels of the tree. Usually, i and j are set to the same value, although i must be greater than or equal to j. Default value is s0.

OPTION 16: The debug option.

d0 Do not print debugging information.

d1 Print debugging information, including how THEO puts together the proof, and what hash codes are assigned to the literals and clauses.

d2 Print extra debugging information including the locations in memory of arrays that are dynamically allocated.

OPTION 17: Print out a long or short version of the proof.

r1 When printing the proof, list only those base clauses used as well as those inferences used.

r0 When printing the proof, list all base clauses and all inferences used.

### 10.3.3 Option to prove a set of theorems

OPTION 18: Prove a set of theorems listed in some file, say PROVEALL

batch Tells THEO that there is a set of theorems in PROVEALL for which proofs are sought. PROVEALL must follow the redirection symbol, "<." At the system prompt, type theo batch t30 < PROVEALL to tell THEO to prove every theorem listed in the file PROVEALL with a time limit of 30 seconds per theorem.

### 10.3.4 Obtaining help by typing "?"

If a "?" appears anywhere on the line naming the theorem, a help menu appears on the screen. After presenting the options available, THEO again requests the name of a theorem.

## 10.4 User Interaction During the Search

User interaction during the search can take place when the search halts after printing a clause. If Options 12 and 13 are set as p6 s6, for example, then when searching, THEO will stop after printing each clause in the tree at a depth of six or less; the user can then tell THEO how to proceed. Typing a:

q tells THEO to quit the search,
k tells THEO to terminate this iteration and go on to the next one,
r tells THEO to terminate this iteration and repeat it,
x tells THEO to terminate the search at this branch,
t tells THEO to print the time consumed,
n tells THEO to print the number of nodes searched,
a tells THEO to print the partially formed proof,
c tells THEO to print the current path under search,
y tells THEO to print the tree structure,
vi tells THEO to print clause i,
+(-) tells THEO to increment (decrement) the step and print depth,
p tells THEO to print the current step and print depth,
d tells THEO to turn on the debug option if it was off and off if it was on,
? tells THEO to print the help menu,
CR"(printing a carriage return) tells THEO to continue with the search.

## 10.5 Examples of User Options

In response to asking for the name of a theorem, type:

STARK036.THM k0 z1

THEO will then attempt to prove STARK036.THM skipping Phase 0 and looking for a linear-nc proof. If you enter:

STARK036.THM p4 s2

THEO will attempt to prove STARK036.THM. It will print all clauses at the first four levels of the search tree during Phases 1 and 2. After printing each clause at the first two levels, THEO will wait until told how to proceed.

## 10.6 The Printout Produced by THEO

When THEO attempts to find a proof, the proof, along with other information, is printed on the screen and stored in a disk file as discussed earlier. This section explains this information using the printout produced when the theorem S44WOS20.THM is proved with default options. Some of the information is printed out before the search begins, some during the search, and still other information after the search is done. The printout is presented here. That produced by the computer appears in bold font. Comments are added to the printout in brackets.

**Theorem:../t/theorems/S44WOS20.THM**

[The name of the theorem is printed out.]

**Predicates: p Equal o**
**Functions: d . e c b a : i f g**

[A listing of predicates, then constants in the negated conclusion followed by a period, then other constants, followed by another period, followed by functions.]

**EQ: 6.7.8**
**ESAF: 13 14 15 22 23**
**ESAP: 10 11 12 19**

[A listing of all equality axioms, all equality substitution axioms for functions, and all equality substitution axioms for predicates.]

**OPT: k1 m1 z0 h1 n1 b1  x12:X6:N1 v12 l3 t3600**

[The options used to prove this theorem: k1 => use eliminate; m1 => search for a linear-merge proof; z0 => do not search for a linear-nc proof; h1 => use clause_hash_table; n1 => use the extended search strategy; b1 => carry out simple subsumption in the search tree; x12 => bound the maximum number of terms in a literal of an inference to 12; X6 => bound the maximum number of terms in a literal in the negated conclusion to 6; v12 => eliminate inferences with more than 12 different variables; l3 => delete inferences with more than three literals; t3600 => a maximum of 3600 seconds are allotted to prove the theorem.]

**0 <BC: 33 NC: 1 AC: 35 U: 16>**

[When Phase 0 finishes, THEO prints the time consumed, the number of base clauses, the number of clauses in the negated conclusion, the number of clauses in the possibly revised base clauses, and the number of unit clauses in the original base clauses. Phase 1 begins now; the following information is printed on the screen and saved on disk as the search progresses.]

```
1 {T1 N355 R928 F29 C604 H2 h0 U3415}
```

[The iteration number followed by the time in seconds, the nodes searched, the binary resolvents formed, the binary factors formed, the number of clauses entered in clause_hash_table, the number of clauses found in clause_hash_table, the number of clauses unable to be entered in clause_hash_table because an empty location cannot be found, and the number of unit clauses.]

```
2 {T2 N2057 R6235 F208 C4069 H729 h0 U20050}
3 {T7 N14064 R42244 F1586 C29814 H6526 h0 U76608}
```

[The same information on the second and third iterations.]

```
.4{T16 N38483 R112087 F3466 C74172 H22877 h0 U107679}*
```

[A proof is found, denoted by the asterisk ("*"). The "." at the beginning of the line and those on lines to come indicate that a line — in this case, the main line — of the proof has been found. THEO now enters Phase 2, searching for the hash codes of the required unit clauses and the other lines.]

```
.1 {T17 N353 R113015 F3495 C604 H2 h0 U3415}
.2 {T17 N574 R113597 F3510 C824 H116 h0 U3658}*
```

[The two "."s at the head of the lines indicate that one line was found on the first iteration and the second on the second iteration. Another search is necessary.]

```
.1 {T18 N94 R113848 F3515 C978 H116 h0 U4005}*
```

[The last line is found! THEO has assembled the proof and enters Phase 3. The proof is then verified and finally printed.]

```
Proof Found!

Axioms:
 1 >pexx
```

```
  2 >pgxxe
  3: pxyfxy
  4 >~pxyz ~pyuv ~pzuw pxvw
  5: ~pxyz ~pyuv ~pxvw pzuw
  6: Equalxx
  7: ~Equalxy Equalyx
  8: ~Equalxy ~Equalyz Equalxz
  9: ~pxyz ~pxyu Equalzu
 10: ~Equalxy ~pzux pzuy
 11: ~Equalxy ~pzxu pzyu
 12: ~Equalxy ~pxzu pyzu
 18 >~ox ~oy ~pxyz oz
 19 >~ox ~Equalxy oy
 20 >~ox ogx
 21: oe
 22: ~Equalxy Equalizxizy
 23: ~Equalxy Equalixziyz
 24 >ox oy oixy
 25 >ox oy pxixyy
 26 >~pxyz ~pxuz Equalyu
 27: ~pxyz ~puyz Equalxu
 28 >Equalggxx
 29 >~oa
 30 >ob
 31 >pbgac
 32 >pacd

Negated conclusion:
33S>~od

Phase 0 clauses used in proof:
34>(30a*20a) ogb

Phases 1 and 2 clauses used in proof:
36S(33a,25b) ox pxixdd
 37S(36b,26a) ox ~pxyd Equalixdy
  38S(37b,32a) oa Equaliadc
   39S(38a,29a) Equaliadc
40S(33a,24b) ox oixd
 41S(40b,19a) ox ~Equalixdy oy
  42S(41a,29a) ~Equaliadx ox
   43S(42a,39a) oc
```

```
44: (20b,19a) ~ox ~Equalgxy oy
 45: (44b,28a) ~ogx ox
  46: (45b,29a) ~oga
47: (31a,4b)  ~pxby ~pygaz pxcz
 48: (47b,1a)  ~pxbe pxcga
  49: (48a,2a)  pgbcga
   50: (49a,18c) ~ogb ~oc oga
    51: (50a,34a) ~oc oga
     52S(51a,43a) oga
      53S(52a,46a) []

PHASE 0:0s, PHASE 1:16s, PHASE 2:2s, Total Time:18s
NOD: 39151  RES: 113848  FAC: 3515 T:12 V:12 L:3
CTE: 74172  CTH: 22877   CTF: 0      CSZ: 2097152
UTE: 107679 UTH: 176531  UTF: 0
BAS: 33     RED: 35      LEN: 1 + 18
OPT: k1 m1 z0 h1 n1 b1 t3600
```

Clauses with an ">" are either base clauses or clauses generated in Phase 0 that are used in the proof. Those that have an "S" following the clause number are descendants of the negated conclusion. The data on the search — the last six lines — indicates that Phase 0 took 0 seconds, Phase 1 took 16 seconds, Phase 2 took 2 seconds, and the total time was 18 seconds. The other information is as follows.

NOD: Number of nodes in the trees searched in Phases 1 and 2.
RES: Number of binary resolvents formed in Phases 1 and 2.
FAC: Number of binary factors formed in Phases 1 and 2.
T: Maximum number of terms in a literal kept during the search.
V: Maximum number of different variables in a clause kept during the search.
L: Maximum number of literals in a clause kept during the search.
CTE: Number of nonunit clauses entered in clause_hash_table in Phase 1.
CTH: Number of times entries in clause_hash_table were "hit" in Phase 1.
CTF: Number of nonunit clauses that caused hash table overflows in Phase 1.
CSZ: Maximum number of entries possible in clause_hash_table.
UTE: Number of unit clauses placed in clause_hash_table in Phase 1.
UTH: Number of unit hash table resolutions in Phase 1.
UTF: Number of unit clauses that caused hash table overflows in Phase 1.
BAS: Number of base clauses.
RED: Number of base clauses after Phase 0.
LEN: Number of inferences used in Phase 0 plus the number of inferences used in Phases 1 and 2 in the proof.
OPT: Settings for the search options (see Section 10.3.1).

## 10.7 A Second Example: The Printout Produced with the d1 Option

The following is the output of THEO when using the d1 option and when proving S50WOS26.THM. When using this option, the lines found during the search and the contents of cl are among the data printed out as the search progresses.

```
S50WOS26.THM

Predicates: q n p Equal
Functions: e d a . c b : h j
EQ: 16.17.18
ESAF:     ESAP: 8 9 11 13 14 15
OPT: k1 m1 z0 h1 n1 b1  x9:X5:N3 v12 13 t3600
   0 <BC: 24 NC: 1 AC: 28 U: 17>
1 {T0 N344 R1117 F27 C943 H13 h0 U2711}
2 {T0 N2153 R7142 F153 C5249 H1164 h0 U5509}
3 {T1 N11102 R37634 F964 C25786 H9199 h0 U17256}
4 {T3 N30666 R109655 F3084 C69361 H34264 h0 U28826} *
```

[The main line of the proof has been found! It involves six binary resolutions and two unit hash table resolutions. The two unit hash table resolutions resolved away literal ~nab in clause 35 and literal qaxb in clause 38. ]

```
SAVE LINE PLY:6 NODES:30666 ITER:4
34: (31a,10b) ~nxb nxd
 35: (34b,27b) <~nab> ~qxxd nax
  36: (35b,2a)  ~qxxd qahaxx
   37: (36b,14b) ~qxxd ~Equalhaxy qayx
    38: (37a,22a) ~Equalhabx <qaxb>
     39: (38a,16a) []
```

[The information from the main line is stored in cl_array as shown.]

```
CL_ARRAY
1592 L   31,1 R   10,2  HC 0
1593 L 1592,2 R   27,2  HC 0
1594 L 1593,1 H -20535345 1622108919  HC 0
1595 L 1594,2 R    2,1  HC 0
1596 L 1595,2 R   14,2  HC 0
1597 L 1596,1 R   22,1  HC 0
1598 L 1597,2 H 708884719 1629164075  HC 0
1599 L 1598,1 R   16,1  HC 0
```

[The hash codes that must be found are placed on hash_code_list. The list header, printed following this comment, says that there are two hash codes to discover and that none have been found yet.]

```
HASH CODE LIST   To Discover[D]: 2  Found[F]: 0
D[0]:-20535345 1622108919  D[1]:708884719 1629164075
```

[Phase 2 begins now with the search for the two hash codes D[0] and D[1].]

```
Hash code 0 found

SAVE LINE PLY:1 NODES:30712 ITER:1
34: (29a,27b) <~qxxd> nax
```

[The clause whose hash code matches D[0] is found. The hash code for the instance of qxxd is the same as base clause: 22 qbbd. This leads to two new clauses being added to cl as shown here.]

```
Hash code same as base clause:22: qbbd

CL_ARRAY
1590 L   29,1 R   27,2  HC 0
1591 L 1590,1 R   22,1  HC 1
1592 L   31,1 R   10,2  HC 0
1593 L 1592,2 R   27,2  HC 0
1594 L 1593,1 R 1591,1  HC 0
1595 L 1594,2 R    2,1  HC 0
1596 L 1595,2 R   14,2  HC 0
1597 L 1596,1 R   22,1  HC 0
1598 L 1597,2 H 708884719 1629164075  HC 0
1599 L 1598,1 R   16,1  HC 0

HASH CODE LIST   To Discover[D]: 1  Found[F]: 1
F[0]:-20535345 1622108919  D[1]:708884719 1629164075

1 {T4 N344 R110774 F3111 C945 H13 h0 U2708}
2 {T4 N2155 R116800 F3237 C5255 H1164 h0 U5506}
3 {T5 N11102 R147292 F4048 C25792 H9199 h0 U17256}

SAVE LINE PLY:6 NODES:55083 ITER:4
34: (32a,6b)  ~qxyz ~qzuv qxjuyv
 35: (34b,22a) ~qxyb qxjbyd
  36: (35b,5a)  ~qxyb ~qxzd Equalzjby
   37S>(36b,24a) ~qaxb Equalejbx
    38S>(37b,8b)  ~qaxb ~nyjbx nye
     39: (38b,33a) ~qaxb <nbe>
```

[The clause whose hash code matches D[1] is found, ~qaxb. The line yielding this clause shows that another pass is necessary to find the unit clause, nbe.]

```
CL_ARRAY
1583 L    32,1 R    6,2  HC 0
1584 L 1583,2 R   22,1  HC 0
1585 L 1584,2 R    5,1  HC 0
1586 L 1585,2 R   24,1  HC 0
1587 L 1586,2 R    8,2  HC 0
1588 L 1587,2 R   33,1  HC 0
1589 L 1588,2 H 948759567 2108141559   HC 2
1590 L    29,1 R   27,2  HC 0
1591 L 1590,1 R   22,1  HC 1
1592 L    31,1 R   10,2  HC 0
1593 L 1592,2 R   27,2  HC 0
1594 L 1593,1 R 1591,1  HC 0
1595 L 1594,2 R    2,1  HC 0
1596 L 1595,2 R   14,2  HC 0
1597 L 1596,1 R   22,1  HC 0
1598 L 1597,2 R 1589,1  HC 0
1599 L 1598,1 R   16,1  HC 0
HASH CODE LIST   To Discover[D]: 0  Found[F]: 2
F[0]:-20535345 1622108919  F[1]:708884719 1629164075

4 {T6 N24418 R195704 F5477 C52446 H28233 h0 U26079} *

HASH CODE LIST   To Discover[D]: 1  Found[F]: 2
F[0]:-20535345 1622108919  F[1]:708884719 1629164075
D[2]:948759567 2108141559
```

[Another pass in Phase 2 is necessary to find the unit clause with the hash code D[2].]

```
1 {T6 N344 R196821 F5504 C53389 H28246 h0 U28790}
2 {T6 N2153 R202846 F5630 C57695 H29397 h0 U31588}

Hash code 2 found

SAVE LINE PLY:5 NODES:62773 ITER:3
34S>(29a,4c)  <~pa> ~qxxd nax
 35S>(34a,22a) nab
  36S>(35a,10a) ~nbx nax
   37S>(36b,27b) ~nbx ~qyyx nay
    38: (37b,23a) ~nbe <nac>

Hash code same as base clause:21: pa
```

[The unit clause whose hash code is D[2] is found, but now two more must be found. One hash code is that of base clause 21: pa, leaving only the unit clause that resolved away nac in clause 38: (37b,23a) ~nbe <nac>to be found.]

```
CL_ARRAY
1576 L    29,1 R    4,3  HC 0
1577 L 1576,1 R   21,1  HC 0
1578 L 1577,1 R   22,1  HC 0
1579 L 1578,1 R   10,1  HC 0
1580 L 1579,2 R   27,2  HC 0
1581 L 1580,2 R   23,1  HC 0
1582 L 1581,2 H 1672636468 -287205839  HC 3
1583 L    32,1 R    6,2  HC 0
1584 L 1583,2 R   22,1  HC 0
1585 L 1584,2 R    5,1  HC 0
1586 L 1585,2 R   24,1  HC 0
1587 L 1586,2 R    8,2  HC 0
1588 L 1587,2 R   33,1  HC 0
1589 L 1588,2 R 1582,1  HC 2
1590 L    29,1 R   27,2  HC 0
1591 L 1590,1 R   22,1  HC 1
1592 L    31,1 R   10,2  HC 0
1593 L 1592,2 R   27,2  HC 0
1594 L 1593,1 R 1591,1  HC 0
1595 L 1594,2 R    2,1  HC 0
1596 L 1595,2 R   14,2  HC 0
1597 L 1596,1 R   22,1  HC 0
1598 L 1597,2 R 1589,1  HC 0
1599 L 1598,1 R   16,1  HC 0
HASH CODE LIST   To Discover[D]: 0  Found[F]: 3
F[0]:-20535345 1622108919  F[1]:708884719 1629164075
F[2]:948759567 2108141559

3 {T7 N6717 R219235 F6062 C68749 H33330 h0 U40695} *

HASH CODE LIST   To Discover[D]: 1  Found[F]: 3
F[0]:-20535345 1622108919  F[1]:708884719 1629164075
F[2]:948759567 2108141559  D[3]:1672636468 -287205839

Hash code 3 found
```

[One more pass in Phase 2 is necessary.]

```
SAVE LINE PLY:2 NODES:61893 ITER:1
34S>(29a,27b) ~qxxd nax
 35: (34b,20a) <~qbbd> ~nac
```

```
Hash code same as base clause:22: qbbd
```

[The unit clause with the hash code being sought has been found. The literal ~qbbd in clause 35 was resolved away by base clause 22 qbbd.]

```
CL_ARRAY
1573 L   29,1 R   27,2  HC 0
1574 L 1573,2 R   20,1  HC 0
1575 L 1574,1 R   22,1  HC 4
1576 L   29,1 R    4,3  HC 0
1577 L 1576,1 R   21,1  HC 0
1578 L 1577,1 R   22,1  HC 0
1579 L 1578,1 R   10,1  HC 0
1580 L 1579,2 R   27,2  HC 0
1581 L 1580,2 R   23,1  HC 0
1582 L 1581,2 R 1575,1  HC 3
1583 L   32,1 R    6,2  HC 0
1584 L 1583,2 R   22,1  HC 0
1585 L 1584,2 R    5,1  HC 0
1586 L 1585,2 R   24,1  HC 0
1587 L 1586,2 R    8,2  HC 0
1588 L 1587,2 R   33,1  HC 0
1589 L 1588,2 R 1582,1  HC 2
1590 L   29,1 R   27,2  HC 0
1591 L 1590,1 R   22,1  HC 1
1592 L   31,1 R   10,2  HC 0
1593 L 1592,2 R   27,2  HC 0
1594 L 1593,1 R 1591,1  HC 0
1595 L 1594,2 R    2,1  HC 0
1596 L 1595,2 R   14,2  HC 0
1597 L 1596,1 R   22,1  HC 0
1598 L 1597,2 R 1589,1  HC 0
1599 L 1598,1 R   16,1  HC 0
HASH CODE LIST   To Discover[D]: 0  Found[F]: 4
F[0]:-20535345 1622108919  F[1]:708884719 1629164075
F[2]:948759567 2108141559  F[3]:1672636468 -287205839

1 {T7 N93 R219555 F6069 C68981 H33330 h0 U42657} *
HASH CODE LIST   To Discover[D]: 0  Found[F]: 4
F[0]:-20535345 1622108919  F[1]:708884719 1629164075
F[2]:948759567 2108141559  F[3]:1672636468 -287205839

Proof Found!
```

[The proof is now complete and is printed out.]

```
Axioms:
1 >qxyjxy
2 >~nxy qxhxyy
3 >~qxyz nxz
4 >~px ~qyyz ~nxz nxy
5 >~qxyz ~qxuz Equaluy
6 >~qxyz ~qyuv ~qzuw qxvw
7: ~qxyz ~qyuv ~qxvw qzuw
8 >~nxy ~Equalzy nxz
9: ~nxy ~Equalxz nzy
10 >~nxy ~nyz nxz
11: ~px ~Equalxy py
12: ~qxyz ~qxyu Equalzu
13: ~Equalxy ~qzux qzuy
14 >~Equalxy ~qzxu qzyu
15: ~Equalxy ~qxzu qyzu
16 >Equalxx
17: ~Equalxy Equalyx
18: ~Equalxy ~Equalyz Equalxz
19 >~qxyz qyxz
20 >~nxb ~nxc
21 >pa
22 >qbbd
23 >qcce

Negated conclusion:
24S>qaed

Phase 0 clauses used in proof:
27 >(21a*4a)  ~qxxy ~nay nax
29S>(24a*3a)  nad
31 >(22a*3a)  nbd
32 >(19a*1a)  qxyjyx
33 >(3a*1a)   nxjxy
Phases 1 and 2 clauses used in proof:
34S>(29a,27b) ~qxxd nax
 35S>(34b,20a) ~qbbd ~nac
  36S>(35a,22a) ~nac
37S>(29a,4c)  ~pa ~qxxd nax
 38S>(37a,21a) ~qxxd nax
  39S>(38a,22a) nab
   40S>(39a,10a) ~nbx nax
    41S>(40b,27b) ~nbx ~qyyx nay
     42S>(41b,23a) ~nbe nac
      43S>(42b,36a) ~nbe
44: (32a,6b)  ~qxyz ~qzuv qxjuyv
 45: (44b,22a) ~qxyb qxjbyd
```

```
  46: (45b,5a)  ~qxyb ~qxzd Equalzjby
   47S>(46b,24a) ~qaxb Equalejbx
    48S>(47b,8b)  ~qaxb ~nyjbx nye
     49S>(48b,33a) ~qaxb nbe
      50S>(49b,43a) ~qaxb
51S>(29a,27b) ~qxxd nax
 52S>(51a,22a) nab
53: (31a,10b) ~nxb nxd
 54: (53b,27b) ~nab ~qxxd nax
  55S>(54a,52a) ~qxxd nax
   56S>(55b,2a)  ~qxxd qahaxx
    57S>(56b,14b) ~qxxd ~Equalhaxy qayx
     58S>(57a,22a) ~Equalhabx qaxb
      59S>(58b,50a) ~Equalhabx
       60S>(59a,16a) []

PHASE 0: 0 s, PHASE 1: 4 s, PHASE 2: 4 s, Total Time: 8 s
NOD: 61894       RES: 219555     FAC: 6069    T: 9 V: 12 L: 3
CTE: 69361       CTH: 34264      CTF: 0       CSZ: 2097152
UTE: 28826       UTH: 100241     UTF: 0
BAS: 24          RED: 28         LEN: 5 + 27
OPT: k1 m1 z0 h1 n1 b1 t3600
```

## Exercises for Chapter 10

10.1. (a) Print the entire search tree for STARK103.THM. (b) How many entries were made in clause_hash_table? (c) Which clauses produced these entries? (Hint: to answer part (c) correctly, it is necessary to examine the source code for THEO in the file search.c. Clauses are added to clause_hash_table by calling add_clause_to_ht from the function install_clause.)

10.2. (a) Which theorem in the directory THEOREMS is the most difficult for THEO? (b) What is this theorem about? (c) For which theorem was the proof the longest? How long? (d) For which theorem was the number of entries in clause_hash_table the most? How many?

10.3. Explain how the proof of STARK087.THM is found by THEO.

10.4. Find a linear proof of STARK017.THM (see Section 5.1) that begins by forming the binary resolvent 13: (12a,7a) less(F(A),F(A)) | ~ prime(F(A)), and then by forming the binary resolvent 14: (13b,8b) less(F(A),F(A)) |

divides(SK(F(A)),F(A)). (Hint: a linear proof exists with seven additional inferences. Use options k0 h0 b0 l10 x20 p2 s2 when calling the theorem, and then use options k, r, and x to guide the search, as described in Section 10.4.)

10.5. Use THEO interactively to confirm the linear-merge-nc proof you found for SQROOT.THM as the solution to Exercise 6.12 is correct.

10.6. In an interactive fashion, THEO can be used to **verify** whether a proof is correct or not. For example, consider the theorem STARK103.THM shown on pages 100–107. Two proofs are shown in Figure 10.15. Are they correct? If not, what are the errors? Describe how you used THEO to decide.

10.7. The options used to prove a theorem greatly affect the time required to find a proof. Consider the four theorems STARK075.THM, STARK118.THM, S39WOS15.THM, and S46WOS22.THM. Try to prove each of them using the eight combinations of the three options (1) linear-nc ON/OFF (z1/z0), (2) clause_hash_table ON/OFF(h1/h0), and (3) extended search strategy ON/OFF(n1/n0). To do this, edit the file PROVEALL, deleting all but these four theorems. Then, create eight copies of each theorem in some file, say PROVEALL2, adding the appropriate options to each one. Can you draw any conclusions about the relative effectiveness of these options?

10.8. Jeff Pelletier (1986) credits Len Schubert for the Aunt Agatha riddle (see AGATHA.WFF). A resident of Dreadsbury Mansion killed Aunt Agatha.

```
Proof 1
15: (14a,6c)  ~SBA | ~SAB
16: (9c,13a)  ~MxA | MxB
18: (7d,13a)  ~MxB | MxA
28: (18a,2b)  SBx | MFBxA
29: (28b,3b)  SBA
30: (29a,15a) ~SAB
31: (16a,2b)  SAx | MFAxB
32: (31b,3b)  SAB
33: (32a,30a) [ ]
```

```
Proof 2
15: (9a,2b)   Sxy | MFxyz | ~UNuxz
16: (15b,3b)  Sxy | ~UNzxy
17: (16b,13a) SAB
18: (14a,6c)  ~SBA | ~SAB
19: (18b,17a) ~SBA
20: (19a,2a)  MFBAB
21: (14a,6c)  ~SBA | ~SAB
22: (21b,17a) ~SBA
23: (22a,3a)  ~MFBAB
24: (23a,7c)  ~MFBAx | MFBAy | ~UNyAx
25: (24c,13a) ~MFBAB | MFBAA
26: (25a,20a) MFBAA
27: (26a,23a) [ ]
```

Figure 10.15. Two different proofs (?) of STARK103.THM.

Agatha, the butler, and Charles live there and are its only residents. Killers hate their victims, and they are never richer than their victims. Charles hates no one that Agatha hates. Agatha hates everyone except the butler. The butler hates everyone not richer than Agatha. The butler hates everyone Agatha hates. No one hates everyone. Agatha is not the butler. Prove: Aunt Agatha killed herself.

10.9. In Exercise 2.8, you were asked to pose as a theorem the problem of finding a solution to the missionaries and cannibals problem. Similarly, in Exercise 2.9, you were asked to pose as a theorem the problem of finding a solution to the eight-puzzle. Now, use THEO to solve these problems.

10.10. Find a proof to the knights and knaves riddle (see KNGTS.THM), from Steve Smullyan's book *What is the Name of this Book?* (see Ohlbach 1985).

Suppose you are an inhabitant of an island of knights and knaves. The knights always tell the truth and the knaves always lie. You fall in love with a girl there and wish to marry her. However, this girl has strange tastes: for some odd reason she does not wish to marry a knight. She wants to marry a knave. But she wants a rich knave, not a poor one. Suppose, in fact, that you are a rich knave. You are allowed to make only one statement. Can you convince her that you are a rich knave?

Let T(x) say "x is true," and Says(x,y) say "x can say y." The theorem can be stated as shown in Figure 10.16.

10.11. Using the set of clauses shown in Figure 10.17, attempt to solve the Lion and the unicorn problem (see LION.THM) as considered by H. J. Ohlbach and M. Schmidt-Schauss in "The lion and the unicorn," J. Autom. Reasoning, 1, 1985, 327–332:

When Alice entered the forest of forgetfulness, she did not forget everything, only certain things. She often forgot her name, and the most likely thing for her to forget was the day of the week. Now, the lion and the unicorn were frequent visitors to this forest. These two are strange creatures. The lion lies on Mondays, Tuesdays, and Wednesdays, and tells the truth on the other days of the week. The unicorn, on the other hand, lies on Thursdays, Fridays, and Saturdays, but tells the truth on the other days of the week. One day, Alice met the lion and the unicorn resting under a tree. They made the following statements: (1) Lion: Yesterday was one of my lying days; (2)Unicorn: Yesterday was one of my lying days.

Alice, a bright girl, was able to deduce the day of the week. What was it?

```
Axioms
 1  ~T(knight(x)) | ~T(knave(x))
 2  T(knight(x)) | T(knave(x))
 3  ~T(rich(x)) | ~T(poor(x))
 4  T(rich(x)) | T(poor(x))
 5  ~T(knight(x)) | ~Says(x,y) | T(y)
 6  ~T(knight(x)) | Says(x,y) | ~T(y)
 7  ~T(knave(x)) | ~Says(x,y) | ~T(y)
 8  ~T(knave(x)) | Says(x,y) | T(y)
 9  ~T(and(x,y)) | T(x)
10  ~T(and(x,y)) | T(y)
11  T(and(x,y)) | ~T(x) | ~T(y)

Negated conclusion
12  ~Says(i,x) | ~T(and(knave(i()),rich(i)))
13  Says(i,x)  | T(and(knave(i()),rich(i)))
```

Figure 10.16. Knights and knaves riddle.

10.12. Len Schubert's steamroller problem is a classic problem for testing theorem provers (see Pelletier 1986).

Given: Wolves, foxes, birds, catepillars, and snails are animals, and there are some of each of them. Also, there are some grains and grains are plants. Every animal either likes to eat all plants or all animals much smaller than itself that like to eat some plants. Caterpillars and snails are much smaller than birds, which are much smaller than foxes, which in turn are much smaller than wolves. Wolves do not eat foxes or grains, while birds like to eat caterpillars but not snails. Caterpillars and snails like to eat some plants.

Prove: There is an animal that likes to eat a grain-eating animal.

First, set up a set of wffs corresponding to the English statements. Then, transform the wffs to clauses using COMPILE. Then, attempt to find a proof using THEO. Use the e1 option to follow the steps carried out by the function eliminate in THEO.

```
Axioms
1: MO(x) | TU(x) | WE(x) | TH(x) | FR(x) | SA(x) | SU(x)

2: ~MO(x) | ~TU(x)
3: ~MO(x) | ~WE(x)
4: ~MO(x) | ~TH(x)
5:~MO(x) | ~FR(x)
6: ~MO(x) | ~SA(x)
7: ~MO(x) | ~SU(x)
8: ~TU(x) | ~WE(x)
9: ~TU(x) | ~TH(x)
10: ~TU(x) | ~FR(x)
11: ~TU(x) | ~SA(x)
12: ~TU(x) | ~SU(x)
13: ~WE(x) | ~TH(x)
14: ~WE(x) | ~FR(x)
15: ~WE(x) | ~SA(x)
16: ~WE(x) | ~SU(x)
17: ~TH(x) | ~FR(x)
18: ~TH(x) | ~SA(x)
19: ~TH(x) | ~SU(x)
20: ~FR(x) | ~SA(x)
21: ~FR(x) | ~SU(x)
22: ~SA(x) | ~SU(x)

23: ~MO(ystday(x)) | TU(x)
24: ~TU(x) | MO(ystday(x))
25: ~TU(ystday(x)) | WE(x)
26: ~WE(x) | TU(ystday(x))
27: ~WE(ystday(x)) | TH(x)
28: ~TH(x) | WE(ystday(x))
29: ~TH(ystday(x)) | FR(x)
30: ~FR(x) | TH(ystday(x))
31: ~FR(ystday(x)) | SA(x)
32: ~SA(x) | FR(ystday(x))
33: ~SA(ystday(x)) | SU(x)
34: ~SU(x) | SA(ystday(x))
35: ~SU(ystday(x)) | MO(x)
36: ~MO(x) | SU(ystday(x))

37: ~Mem(x, lydays(lion)) | MO(x) | TU(x) | WE(x)
38: ~Mem(x, lydays(unicorn)) | TH(x) | FR(x) | SA(x)
39: ~MO(x) | Mem(x,lydays(lion))
40: ~TU(x) | Mem(x,lydays(lion))
41: ~WE(x) | Mem(x,lydays(lion))
42: ~TH(x) | Mem(x, lydays(unicorn))
43: ~FR(x) | Mem(x, lydays(unicorn))
44: ~SA(x) | Mem(x, lydays(unicorn))
45: Mem(x, lydays(z)) | ~LA(z,x,y) | Mem(y, lydays(z))
46: Mem(x, lydays(z)) | LA(z,x,y) | ~Mem(y,lydays(z))
47: ~Mem(x, lydays(z)) | ~LA(z,x,y) | ~Mem(y, lydays(z))
48: ~Mem(x, lydays(z)) | LA(z,x,y) | Mem(y, lydays(z))
49: LA(lion, today, ystday(today))
50: LA(unicorn, today, ystday(today))

Negated conclusion
51: ~TH(today)
```

Figure 10.17. The lion and the unicorn problem.

# 11 A Look at the Source Code of HERBY

This chapter introduces you to the source code of HERBY. Section 11.1 briefly describes the source files. Section 11.2, and, in particular, Figure 11.1, describe how the functions in these files are linked. A brief description of the main functions in HERBY is presented in Section 11.3. HERBY represents a theorem at the machine code level as described in Section 11.4. Major array in HERBY are described in Section 11.5.

## 11.1 Source Files for HERBY

The source code for HERBY can be found in HERBYSC and consists of 11 -.c files:

| | |
|---|---|
| herby.c | contains main |
| build.c | supervises building the semantic tree |
| infer.c | generates inferences |
| hgen.c | constructs clauses |
| hhash.c | calculates hash codes for clauses |
| unify.c | checks to see if two literals unify |
| equal.c | functions related to the Equal predicate |
| atom.c | generates atoms using various heuristics |
| tptpi.c | transforms TPTP Problem Library format to HERBY/THEO format |
| affc.c | transforms theorem to machine code representation |
| hdump.c | prints clauses |

and two -.h files:

| | |
|---|---|
| clause.h | data structure definitions |
| const.h | constants related to data structures |

| atom_sel_heuristic(a1) | ash0(a2) | | | |
|---|---|---|---|---|
| | | **grab_number(m4)** | | |
| | | **calc_numb_var_in_lit(hh3)** | | |
| | | copy_atom_from_clause(g5) | **clear_var_substitution_tble(i6)** | |
| | | | [gen_lit(g4)] | |
| | | | **calc_clause_hashcode(hh4)** | |
| | | **install_hb_clause(b3)** | | |
| | | copy_gnd_atom_from_cls(g8) | **clear_var_substitution_tble(i6)** | |
| | | | gen_ground_lit(g7) | **emit_ground_lit(g6)** |
| | | | | **calc_lit_hashcode(hh2)** |
| | | | **calc_clause_hashcode(hh4)** | |
| | ash1(a3) | common_instance(i3) | **standardize(i5)** | |
| | | | **clear_var_substitution_tble(i5)** | |
| | | | [unify(u2)] | |
| | | | gen_unified_lit(g10) | [gen_lit(g4)] |
| | | | | **calc_clause_hashcode(hh4)** |
| | | install_ash_clause(b4) | **calc_numb_var_in_lit(hh3)** | |
| | | | [copy_gnd_atom_from_cls(g8)] | |
| | | | **copy_atom_from_clause(g5)** | |
| | ash2(a4) | binary_resolvent2(i2) | **standardize(i5)** | |
| | | | **clear_var_substitution_tble(i5)** | |
| | | | [unify(u2)] | |
| | | | generate_clause2(g2) | [gen_lit(g4)] |
| | | | | **calc_clause_hashcode(hh4)** |
| | | [common_instance(i3)] | | |
| | | [install_ash_clause(b4)] | | |
| | ash3(a5) | [binary_resolvent2(i2)] | | |
| | | [common_instance(i3)] | | |
| | | [install_ash_clause(b4)] | | |
| | ash4(a6) | [binary_resolvent2(i2)] | | |
| | | [common_instance(i3)] | | |
| | | [install_ash_clause(b4)] | | |
| | | install_unit_and_test(b5) | [copy_clause(g9)] | |
| | | | [binary_resolvents2(i2)] | |
| | | | [install_ash_clause(b4)] | |
| | | | **clear_var_substitution_tble(i5)** | |
| | | | [unify(u2)] | |
| | | | [gen_unified_lit(g10)] | |
| | ash5a(a7), ash5c(a9) | **calc_numb_var_in_lit(hh3)** | | |
| | | [copy_atom_from_clause(g5)] | | |
| | | [copy_gnd_atom_from_cls(g8)] | | |
| | | **install_hb_clause(b3)** | | |
| | ash5b(a8) | [copy_atom_from_clause(g5)] | | |
| | | [copy_gnd_atom_from_cls(g8)] | | |
| | | **install_hb_clause(b3)** | | |

Figure 11.1. HERBY: Function linkage (continued).

| main(m6) | enter_theorem_name(m7) | init_options(m3) | **grab_number(m4)** | Note 1: The functions fatal, time_now, | File Abbreviations |
|---|---|---|---|---|---|
| | CompileTptp(tp12) | | | and print_clause are called by many | |
| | compile_clauses(cc28) | | | functions, but to make this table | m = herby.c |
| | **init_search_space(b1)** | | | clearer, the calls are not shown. | b = build.c |
| | prepare_clauses(m12) | **get_pred_fxn_symbols(m13)** | | Note 2: functions in the files affc.c and | g = hgen.c |
| | | **read_buffer_word(m8)** | | tptpi.c are not shown. They are only | l = infer.c |
| | | format_clause(m14) | **read_buffer_word(m8)** | concerned with reading in the theorem. | hh = hhash.c |
| | | max_term_in_any_lit(m16) | **count_lit_characters(m17)** | Note 3: Functions in bold characters | u = unify.c |
| | | sort_constants_fxns(m10) | **is_fxn_a_constant(m9)** | call no other functions except | e = equal.c |
| | | | **interchange_fxn_const(m11)** | fatal, time_now, and print_clause. | tp = tptpi.c |
| | | **equal_predicate(e1)** | | Note 4: Functions enclosed by square | cc = affc.c |
| | | esa_for_fxns(e2) | **test_equality_fxn(e3)** | brackets call other functions. These | a = atom.c |
| | | esa_for_lits(e4) | **test_lits(e5)** | other functions appear elsewhere in | p = hdump.c |
| | **init_random_hash_array(hh1)** | | | this table. | |
| | calculate_hashcodes(m15) | **calc_lit_hashcode(hh2)** | | Note 5: search_tree, eliminate, | Note: m10 denotes the |
| | | **calc_clause_hashcode(hh4)** | | and install_clause each call many | 10th function in herby.c |
| | **clear_fail_pair_info(m2)** | | | functions. They appear in the linkage | |
| | resolution_stage(a10) | binary_resolvents(i1) | **standardize(i5)** | also in the leftmost column where | |
| | | | **clear_var_substitution_tble(i5)** | the functions they call are shown. | |
| | | | unify(u2) | **occurs(u2)** | |
| | | | | **pop_stack(u3)** | |
| | | | hbgen_clause(g11) | [gen_lit(g4)] | |
| | | | | **calc_clause_hashcode(hh4)** | |
| | | | generate_clause(g1) | [gen_lit(g4)] | |
| | | | | **calc_clause_hashcode(hh4)** | |
| | | | install_clause(b2) | **simple_subsume (b9)** | |
| | supervise_build_tree(m18) | build_tree(b7) | **save_failure_pair(b8)** | | |
| | | | **print_clause(p2)** | | |
| | | | atom_sel_heuristic(a1) | | |
| | | | calc_inferences(b6) | [binary_resolvents(i1)] | |
| | | | | **simple_subsume(b9)** | |
| | add_clause(m20) | **standardize(i5)** | | | |
| | | [unify(u2)] | | | |
| | | [generate_clause(g1)] | | | |
| | | copy_clause(g9) | **clear_var_substitution_tble(i6)** | | |
| | | | gen_lit(g4) | **emit_lit(g3)** | |
| | | | | **calc_lit_hashcode(hh2)** | |
| | | | **calc_clause_hashcode(hh4)** | | |
| | | binary_factors(i4) | **clear_var_substitution_tble(i6)** | | |
| | | | [unify(u2)] | | |
| | | | [generate_clause(g1)] | | |
| | **write_out_results(m19)** | | | | |

Figure 11.1. HERBY: Function linkage.

A makefile is included with the source files.There is considerable overlap of code in the two programs HERBY and THEO. Several files are essentially identical in both programs: affc.c, tptpi.c, equal.c, and unify.c. The two header files are the identical as well.

## 11.2 Function Linkage in HERBY

Figure 11.1 shows the linkage of functions in HERBY, omitting links to functions in the files tptpi.c and affc.c. These functions need not be understood by the user; they are exclusively concerned with reading clauses into HERBY and transforming their text representation into the computer's internal representation, which is described in this chapter. These files have nothing to do with HERBY's effort to find a proof.

Functions in the source files are abbreviated by one or two letters, as shown in the upper right-hand corner of the figure. For example, m10 denotes the 10th function in herby.c. Note that the function atom_sel_heuristics(a1) is the root of a large tree of functions.

## 11.3 A Brief Description of the Main Functions in HERBY

The function main(m6):

1. first determines what theorem to prove, then reads in the theorem using TPTP Problem Library format if so instructed, then transforms the clauses into their internal representation while extracting information about the theorem and setting up arrays:

    enter_theorem_name(m7)
    CompileTptp(ct12)
    compile_clauses(cc28)
    init_search_space(b1)
    prepare_clauses(m12)
    init_random_hash_array(hh1)
    calculate_hash_codes(m15)
    clear_failure_pair_info(m2)

2. then calls resolution_stage(a9), which carries out a preliminary effort to generate clauses to add to the base clauses that might help to find a closed semantic tree (see Section 7.2.2).

```
resolution_stage(a9)
```

3. then attempts to construct a closed semantic tree by first calling:

```
supervise_build_tree(m18)
```

which, in turn, calls:

```
build_tree(b7)
```

which tries to construct a closed semantic tree, generating nodes in depth-first order. Atoms are selected by:

```
atom_sel_heuristics(a1)
```

4. then, finally, if a closed semantic tree is constructed, and if the user wishes to transform the tree to a resolution–refutation proof:

```
add_clause(m20)
```

is repeatedly called after new smaller semantic trees are constructed using the newly added clauses.

## 11.4 Machine Code Representation of a Clause in HERBY and THEO

HERBY and THEO store clauses as illustrated by CLAUSEREP.THM in Figure 11.2. There are two predicates, two functions, and three constants in the theorem. They are assigned integers as follows: P = 1, Q = 2; f = 5, a = 2, g = 4, c = 3, b = 1. The three constants a, b, and c are considered functions of zero arguments. Each clause is preceded by a clause header, and each literal is preceded by a literal header as shown later. zero arguments. Each clause is preceded by a clause header, and each literal is preceded by a literal header as shown later.

```
Axioms
1 P(x,f(a)) | Q(y,y,x)
2 ~Q(f(g(x,c),y,z)

Negated conclusion
3 ~P(g(b,f(x)),y)
```

Figure 11.2. CLAUSEREP.THM.

### 11.4.1 The clause header

Information about a clause is stored in the clause header, as show on the following page.

c_id — contains the integer value representing the position of the clause on clist.

*c_end — is a pointer to the memory location following the last word of the clause.

c_bits — uses individual bits to describe the clause. These bits specify whether the clause is a UNIT clause, a MERGE clause, a BASE clause, an NC clause (a clause from the negated conclusion), or an EQCLS clause (a clause which is one of the equality axioms or one of the equality substi tution axioms). A DEAD clause is a clause added during the process of constructing a semantic tree but not needed for one reason or another (used only by HERBY). A clause is either ACTIVE or not (used only by THEO). It can be inactive for several reasons. Clauses that are not ACTIVE are of no use in obtaining a proof. The LIST bit indicates whether the clause is part of the resolution–refutation proof; it is not used by HERBY. VID indicates whether the variables in the clause are represented by integers 1, 2, 3, ... or by the integers 201, 202, 203, ... . Variable names in two different clauses must be numbered differently before the clauses are resolved together.

c_X, c_Y — contain the 64-bit hash code of the clause. The hash function in HERBY and THEO is the same.

*c_lpar , *c_rpar — are pointers to the clause's left-hand and right-hand parents. If the clause is a base clause, it has no parents. If it a binary factor, it only has a left-hand parent.

c_indices — is used to represent the indices of the resolved literals or factored literals. The first index is stored in the least significant six bits of c_indices, and the second index is stored in the next six bits.

*c_lit — is a pointer to the location of the header of the first literal of the clause.

c_nlit — contains the number of literals in the clause.

c_nv — contains the number of distinct variables in the clause.

### 11.4.2 The literal header

Each literal has a four-word header:

*l_next the first word of the header is a pointer to the next literal.

l_X , l_Y contain the 64-bit hash code of the literal.

l_id stores some information about the literal in bit form. The most significant bit indicates whether the literal is negated or not. This is shown in Figure 11.3.

31 30 29 28 27 26 25 24 23 22 21 20 19 18 17 16 15 14 13 12 11 10 09 08 07 06 05 04 –00

bit 31: Indicates whether literal is negated or not. (NEGMASK)

bit 30: Indicates whether literal has been resolved away but kept in clause as "resolved-away" literal. (ELIMLIT)

bits 29 – 05: Name of predicate, function, or variable.

bits 04 – 00: Indicates the number of arguments of the predicate or function.

Figure 11.3. Meaning of l_id field.

### 11.4.3 Representing the terms of a literal

One word is used for each term of a literal. A variable is represented as follows: x is represented by the (hexadecimal) word 00000001, y by the word 00000002, and so on. A function is represented as follows: bit 31 set to a 1, the name of the function occupies bits 5–30, and the number of arguments occupies bits 0–4. Thus, for some arbitrary theorem, the third function with two arguments would be represented by 80000062.

### 11.4.4 An example of the representation

The exact machine code representation of the clauses of CLAUSEREP.THM is presented here. Let us assume the clauses are stored beginning in memory location 00005000 (hexadecimal).

------------ First clause: P(x,f(a)) | Q(y,y,x) -------------

Location Contents Field or clause symbol; comment

[clause header]

```
5000    00000001   c_id       ; first clause on clist
5001    00005019   *c_end     ; pointer to next clause
5002    00000008   c_bits     ; a base clause
5003    random1    c_X        ; 64-bit hash code of first clause
5004    random2    c_Y        ; second 32 bits of hash code
5005    00000000   *c_lpar    ; there are no parent clauses
5006    00000000   *c_rpar
5007    00000000   c_indices
5008    0000500B   *c_lit     ; pointer to first literal
5009    00000002   c_nlit     ; two literals in the clause
500A    00000002   c_nv       ; two distinct variables, x,y
```

[first literal header, including name and number of arguments of literal]

```
500B    00005012   *l_next    ; pointer to second literal
500C    random3    l_X        ; hash code of first literal
500D    random4    l_Y
500E    00000022   l_id       ; P( , ), first predicate, two arguments
```

[arguments of first literal of first clause]

```
500F    00000001   x          ; first variable
5010    800000A1   f          ; bits 0–4 specify one argument
5011    80000040   a          ; second function, no arguments
```

[second literal header, including name and number of arguments of literal]

```
5012    00000000   *l_next    ; no more literals
```

```
5013    random5     l_X         ; hash code of second literal
5014    random6     l_Y
5015    00000043    l_id        ; Q( , , , ), second predicate, three argu-
                                ; ments
```

[arguments of second literal of first clause]

```
5016    00000002    y           ; second variable
5017    00000002    y           ; second variable
5018    00000001    x           ; first variable
```

------------ Second clause: ~Q(f(g(x,c),y,z) ---------------

```
5019    00000002    c_id        ; second clause on clist
501A    0000502E    *c_end      ; pointer to next clause
501B    00000009    c_bits      ; a BASE clause and UNIT clause
501C    random7     c_X         ; 64-bit hash code of second clause
501D    random8     c_Y         ; second 32 bits of hash code
501E    00000000    *c_lpar     ; there are no parent clauses
501F    00000000    *c_rpar
5020    00000000    c_indices
5021    00005022    *c_lit      ; location of first literal
5022    00000001    c_nlit
5023    00000003    c_nv        ; three distinct variables, x, y, and z
```

[first literal header including name and number of arguments of literal]

```
5024    00000000    *l_next     ; pointer to second literal
5025    random7     l_X         ; hash code of first literal
5026    random8     l_Y
5027    80000043    l_id        ; negated second predicate, three ar-
                                ; guments
```

[arguments of first literal of second clause]

```
5028    800000A1    f           ; fifth function, one argument
5029    80000082    g           ; first five bits specify one argument
502A    00000001    x           ; first variable
502B    80000060    c           ; third function, no arguments
502C    00000002    y           ; second variable
502D    00000003    z           ; third variable
```

--------------- Third clause: ~P(g(b,f(x)),y) -------------

| | | | |
|---|---|---|---|
| 502E | 00000003 | c_id | ; third clause on clist |
| 502F | 00005042 | *c_end | ; pointer to next clause |
| 5030 | 00000109 | c_bits | ; a BASE, UNIT, and NC clause |
| 5031 | random9 | c_X | ; 64-bit hash code of third clause |
| 5032 | random10 | c_Y | ; second 32 bits of hash code |
| 5033 | 00000000 | *c_lpar | ; there are no parent clauses |
| 5034 | 00000000 | *c_rpar | |
| 5035 | 00000000 | c_indices | |
| 5036 | 00005039 | *c_lit | ; location of first literal |
| 5037 | 00000001 | c_nlit | ; one literal in clause |
| 5038 | 00000002 | c_nv | ; two distinct variables, x and y |

[first literal header including name and number of arguments of literal]

| | | | |
|---|---|---|---|
| 5039 | 00000000 | *l_next | ; pointer to second literal |
| 503A | random9 | l_X | ; hash code of first literal |
| 503B | random10 | l_Y | |
| 503C | 80000022 | l_id | ; negated first predicate, two argu-<br>; ments |

[arguments of first literal of third clause]

| | | | |
|---|---|---|---|
| 503D | 80000082 | g | ; fourth function, two arguments |
| 503E | 80000020 | b | ; fifth function, no arguments |
| 503F | 800000A1 | f | ; first function, one arguments |
| 5040 | 00000001 | x | ; first variable |
| 5041 | 00000002 | y | ; second variable |

Note that the hash code of a unit clause is the same as the hash code of its only literal.

## 11.5 Major Arrays in HERBY

There are several major arrays in HERBY. The three main ones are cspace, hb_space, and unit_space. Clauses are stored in the array cspace. clist is a list of pointers, with clist[i] pointing to the (i + 1)th clause in cspace. Atoms are stored in the array hb_space, with hb_list[i] pointing to the (i + 1)st atom in hb_space. Unit clauses are stored in unit_space, with unit_list[i] pointing to the (i+1)st unit in unit_space. Clauses are placed in unit_space by ASH4. These arrays are of the following sizes:

| Space | Size | Pointers to clauses | # of pointers |
|---|---|---|---|
| clause_space | 1500000 int | clist[] | 1000 |
| hb_space | 40000 int | hb_list[] | 200 |
| unit_space | 300000 int | unit_list[] | 1500 |

Bounds on the number of clauses on these three lists are 1000, 200, and 1500, as shown. For very large theorems, these bounds may be too small.

In addition to these three arrays, there is an array, rand_array, where random numbers are initially placed and used by the hash function to assign hash codes to clauses, as described in Sections 9.10.1 and 9.10.2.

The array depth[HSPACE_CLAUSES] has the typedef shown here and is used to store pointers to the clauses on c_list at each level in the semantic tree. Note that depth[0].d_bc is the index of the first base clause on clist and is zero. For THEOREM_11.2.THM, the value of depth[0].d_ec is two because the theorem's three base clauses occupy locations 0, 1, and 2 on clist.

```
typedef struct {
    int d_ec;            ; top clause at this level
    clause *d_rc;        ; pointer to first clause at this level
    int d_tc;            ; clause currently under search at this level
    int d_bc;}           ; bottom clause at this level
    dnode;
```

The array branch[HSPACE_CLAUSES] is an array of integers. When HERBY is trying to find a closed left (right) tree of a node at level k in the semantic tree, branch[k] is a 1 (2).

The array fail[HSPACE_CLAUSES] is an array of integers. When a node at level k fails, fail[k] is set to a value of 2.

The array fail_count[HSPACE_CLAUSES] is an array of integers. fail_count[k] records the number of times an atom at level k resolved away a literal of a clause that failed. If when the construction backs up on a left branch to a node at level k at which the value of fail_count[k] is zero, then the right branch need not be constructed (See Section 7.2.5).

## Exercises for Chapter 11

11.1. Express the binary resolvent of clauses 1 and 2 in CLAUSEREP.THM in machine code format. Assume the first word of the resolvent is placed in location 6000 (hexadecimal).

11.2. Express the binary resolvent of clauses 1 and 2 in CLAUSEREP.THM in machine code format. Assume the first word of the resolvent is placed in location 6000 (hexadecimal).

11.3. Examine the source code for the function standardize. Explain what it does.

11.4. Examine the source code for the function clear_var_substitution_tble. Explain what it does.

11.5. Examine the source code for the function unify and the two functions called by unify, occurs and pop_stack. Explain what these functions do and how. (Note: the code is the most difficult to understand in the entire program!)

11.6. Examine the source code for the function generate_clause and the functions called by it. Explain how they all work.

11.7. Examine the source code of HERBY, and discuss in what way the program makes use of hash codes calculated for clauses.

# 12 A Look at the Source Code of THEO

This chapter introduces you to the source code of THEO. The source files are listed and briefly described in Section 12.1. Section 12.2, and, in particular, Figure 12.1, describe how the function in these files are linked. A brief description of the main functions used by THEO is given in Section 12.3. Section 12.4 reminds the reader that both HERBY and THEO represent a clause with the same machine code format. Section 12.5 considers the major arrays in THEO. Sections 12.6 and 12.7 discuss functions used by THEO related to hashing clauses and reconstructing a proof.

## 12.1 Source Files for THEO

The source code for THEO can be found in the file THEOSC and consists of 12 -.c files:

| | |
|---|---|
| theo.c | contains main |
| search.c | controls search of the resolution–refutation tree |
| infer.c | generates inferences |
| gen.c | constructs clauses |
| hash.c | manages hash table and hash codes |
| unify.c | checks to see if two literals unify |
| pass2.c | coordinates the piecing together of a proof |
| equal.c | functions related to the equality predicate |
| elim.c | attempts to simplify the clauses before the search |
| tptpi.c | transforms TPTP Problems Library format to HERBY/THEO format |
| affc.c | transforms theorem to machine code representation |
| dump.c | prints clauses and other information |

and two -.h files:

| | |
|---|---|
| const.h | contains constants related to the size of the theorem |
| clause.h | contains data structure definitions |

A makefile is included with the source files. Again, you might note that six of these files are the same as those in HERBY.

## 12.2 Function Linkage in THEO

The function linkage in THEO is shown in Figure 12.1, omitting those functions in the files tptpi.c and affc.c. This was also done in Chapter 11 when describing HERBY's code. These two files need not be understood by the user. Both are exclusively concerned with reading clauses into THEO and transforming their text representation into the computer's representation, as discussed in this chapter. They have nothing to do with THEO's effort to find a proof. In the material that follows, the functions in the source files are abbreviated by one or two letters, as shown in the upper right-hand corner of Figure 12.1. For example, m10 denotes the 10th function in the file theo.c. Note that three functions, install_clause(s1), search_tree(s3), and eliminate(el19), are at the root of large trees of functions.

## 12.3 A Brief Description of the Main Functions in THEO

The function main(m24):

1. first determines what theorem to prove, then reads in the theorem using TPTP Problem Library format if so instructed, then transforms the clauses into their internal representation:

enter_theorem_name(m7)
CompileTptp(ct12),
compile_clauses(cc28)

2. then extracts information about the theorem and sets up arrays for storing information:

init_cl_array(ps1)
init_search_space(m7)
init_stats(m2)
prepare_clauses(m19)
init_random_hash_array(m9)
hash_all_base_clauses(m20)

3. then carries out a preliminary effort to simplify the clauses:

| main(m24) | enter_theorem_name(m6) | **CompareStrings(tp6)** | | |
|---|---|---|---|---|
| | | init_options(m3) | **grab_number(m4)** | |
| | CompileTptp(tp12) | | | |
| | compile_clauses(cc28) | | | |
| | **init_cl_array(ps1)** | | | |
| | init_search_space(m7) | **dump_calloc_info(pr6)** | | |
| | **init_stats(m2)** | | | |
| | prepare_clauses(m19) | **get_pred_fxn_symbols(m18)** | | |
| | | **read_buffer_word(m11)** | | |
| | | format_clause(m15) | **read_buffer_word(m11)** | |
| | | calc_max_term_in_any_lit(m17) | **count_lit_characters(m16)** | |
| | | sort_constants_fxns(m14) | **is_fxn_a_constant(m13)** | |
| | | | **interchange_fxn_const(m12)** | |
| | | **equal_predicate(e1)** | | |
| | | esa_for_lits(e4) | **test_lits(e5)** | |
| | | esa_for_fxns(e2) | **test_equality_fxn(e3)** | |
| | init_random_hash_array(m9) | **dump_calloc_info(pr6)** | | |
| | hash_all_base_clauses(m20) | **calc_lit_hc(h6)** | | |
| | | **calc_clause_hashcode(h13)** | | |
| | | add_clause_to_ht(h2) | **contradiction(h3)** | |
| | | calc_alt_hashcodes(h9) | **calc_alt_hash(h12)** | |
| | | | [add_clause_to_ht(h2)] | |
| | | normalize_literal(h7) | **clear_var_substitution_tble(i4)** | |
| | | | **count_vars(h8)** | |
| | eliminate(el19) | | | |
| | iterative_deepening(m21) | init_clause_hash_table(m8) | **dump_calloc_info(pr6)** | |
| | | [add_clause_to_ht(h2)] | | |
| | | **calc_clause_hashcode(h13)** | | |
| | | search_tree(s3) | | |
| | | save_line(ps2) | **delete_found_hcs(ps12)** | |
| | | | delete_alt_hcs(ps13) | **delete_found_hcs(ps12)** |
| | | | | **calc_alt_hash(h12)** |
| | | | elim_base_clause_hc(ps3) | **test_for_base_match(ps5)** |
| | | | | **calc_alt_hash(h12)** |
| | | | | **test_equality(g5)** |
| | | | | [normalize_literal(h7)] |
| | | | elim_same_cont_hc(ps4) | **test_for_cont_match(ps6)** |
| | | | | **calc_alt_hash(h12)** |
| | | | | [normalize_literal(h7)] |
| | | | | **test_for_base_match(ps5)** |
| | | | **move_down_renumber(ps14)** | |
| | | | print_cl_array(ps9) | **print_hash_code_list(ps8)** |
| | | [normalize_literal(h7)] | | |
| | | **print_continuation(pr5)** | | |
| | stack_hash_codes(ps7) | **print_hash_code_list(ps8)** | | |
| | **rearrange_cl_array(ps15)** | | | |

Note 1: The functions fatal, time_now, and print_clause are called by many functions, but to make this table clearer, the calls are not shown.
Note 2: functions in the files affc.c and tptpi.c are not shown. They are only concerned with reading in the theorem.
Note 3: Functions in bold characters call no other functions except fatal, time_now, and print_clause.
Note 4: Functions enclosed by square brackets call other functions. These other functions are shown elsewhere in this table.
Note 5: Functions search_tree, eliminate, install_clause, and generate_clause each call many functions. They appear in the linkage and also in the leftmost column, where the functions they call are shown.

File Abbreviations

m = theo.c
s = search.c
I = infer.c
h = hash.c
ps = pass2.c
el = eliminate.c
u = unify.c
e = equal.c
cc = affc.c
tp = tptpi.c
pr = dump.c
g = gen.c

Note: m10 denotes the 10th function in herby.c

Figure 12.1. THEO: Function Linkage.

| | | | | |
|---|---|---|---|---|
| | search_proof_line(ps20) | resolve_proof_line(ps19) | bres_proof_line(ps16) | **standardize(i3)** |
| | | | | **clear_var_substitution_tble(i4)** |
| | | | | [unify(u2)] |
| | | | | generate_clause(g1) |
| | | | | **install_proof_line(ps18)** |
| | | | | print_cl_array(ps9) |
| | | | factor_proof_line(ps17) | **clear_var_substitution_tble(i4)** |
| | | | | [unify(u2)] |
| | | | | generate_clause(g1) |
| | | | | **install_proof_line(ps18)** |
| | write_out_results(m23) | **get_clause_stats(m22)** | | |
| | **free_space_close_file(m10)** | | | |
| search_tree(s3) | calc_inferences(s2) | binary_resolvents(i1) | **standardize(i3)** | |
| | | | **clear_var_substitution_tble(i4)** | |
| | | | unify(u2) | **occurs(u1)** |
| | | | | **pop_stack(u3)** |
| | | | generate_clause(g1) | |
| | | | install_clause(s1) | |
| | | binary_factors(i2) | **clear_var_substitution_tble(i4)** | |
| | | | [unify(u2)] | |
| | | | generate_clause(g1) | |
| | | | install_clause(s1) | |
| | | **is_unit_in_ht(h4)** | | |
| | | **simple_subsume2(s5)** | | |
| | **advance_to_successor(s4)** | | | |
| | halted_search_options(pr4) | **print_continuation(pr5)** | | |
| | | print_cl_array(ps9) | **print_hash_code_list(ps8)** | |
| | | **print_hash_code_list(ps8)** | | |
| | | **grab_number(m4)** | | |
| | | **help_halted_search(pr3)** | | |
| generate_clause(g1) | generate_literal(g3) | **emit_literal(g2)** | | |
| | | **calc_lit_hc(h6)** | | |
| | **calc_clause_hashcode(h13)** | | | |
| | **test_equality(g5)** | | | |
| install_clause(s1) | **simple_subsume2(s5)** | | | |
| | **test_equality(g5)** | | | |
| | repair_literal(g4) | **calc_clause_hashcode(h13)** | | |
| | [add_clause_to_ht(h2)] | | | |
| | **is_clause_in_ht(h1)** | | | |
| | **is_comp_in_hash(h5)** | | | |
| | **calc_clause_hashcode(h13)** | | | |
| | [calc_alt_hashcodes(h9)] | | | |
| | **on_hash_code_list(ps10)** | | | |

Figure 12.1. THEO: Function Linkage (continued).

| | | | | |
|---|---|---|---|---|
| | alt_on_hash_code_list(ps11) | **on_hash_code_list(ps10)** | | |
| | | **calc_alt_hash(h12)** | | |
| | **print_continuation(pr5)** | | | |
| | **is_test_in_ht(h11)** | | | |
| | test_alt_hashcodes(h10) | **calc_alt_hash(h12)** | | |
| | | **is_test_in_ht(h11)** | | |
| | [normalize_literal(h7)] | | | |
| eliminate(el19) | res_ply_1(el2) | bres_ply_1(el1) | **standardize(i3)** | |
| | | | **clear_var_substitution_tble(i4)** | |
| | | | [unify(u2)] | |
| | | | [generate_clause(g1)] | |
| | | | **simple_subsume2(s5)** | |
| | uncomple_lit(el3) | **delete_clause(el15)** | | |
| | base_subsume(el4) | **simple_subsume(el14)** | | |
| | | **delete_clause(el15)** | | |
| | sucessor_subsume(el5) | add_clause(el16) | bres_replace(el17) | **standardize(i3)** |
| | | | | **clear_var_substitution_tble(i4)** |
| | | | | [unify(u2)] |
| | | | | [generate_clause(g1)] |
| | | | | install_clause(s1) |
| | | | **add_clause_to_ht(h2)** | |
| | | | **test_equality(g5)** | |
| | | **delete_clause(el15)** | | |
| | elim_link_lits(el9) | **det_min_link_lits(el11)** | | |
| | | add_link_resolvents(el12) | bres_link(el18) | **standardize(i3)** |
| | | | | **clear_var_substitution_tble(i4)** |
| | | | | [unify(u2)] |
| | | | | [generate_clause(g1)] |
| | | | | install_clause(s1) |
| | | | **add_clause_to_ht(h2)** | |
| | | **delete_link_clauses(el13)** | | |
| | **elim_long_cls(el10)** | | | |
| | gen_subsumption(el6) | **simple_subsume(el14)** | | |
| | | **delete_clause(el15)** | | |
| | unit_resolvent(el8) | [add_clause(el16)] | | |
| | single_resolvent(el7) | [add_clause(el16)] | | |
| | | **delete_clause(el15)** | | |

Figure 12.1. THEO: Function Linkage (continued).

```
eliminate(el19)
```

4. then performs an iteratively deepening depth-first search looking for a resolution–refutation proof:

```
prove_by_iterative_deepening(m21)
```

which calls:

```
search_tree(s3)
save_line(ps2)
```

where search_tree(s3) controls the generation of clauses in the search tree. save_line(ps2) supervises the saving of lines of the proof as they are found.

5. then after a proof has been found, calls:

```
stack_hash_codes(ps7)
```

and

```
rearrange_cl_array(ps15)
```

to help construct the proof from the pieces that have been found.

## 12.4 Machine Code Representation of a Clause in THEO

As discussed in Chapter 11, THEO represents clauses in the same format as is done by HERBY.

## 12.5 Major Arrays in THEO

There are several major arrays in THEO. Clauses are stored in the array cspace. clist is a list of pointers with clist[i] pointing to the (i+1)th clause in cspace. Elements of the array depth are of the typedef dnode shown here and are used to store indices of clauses and pointers to clauses at each level in the search tree. Note that depth[0].d_bc is a pointer to the first base clause.

```
typedef struct {
        int      d_ec;          ; top clause at this level
        clause   *d_rc;         ; pointer to first clause at this level
        int      d_tc;          ; clause under search at this level
        int      d_bc;}         ; bottom clause at this level
                 dnode;
```

So far, everything said in this section is also true for HERBY.

Hash codes representing clauses and information about the clauses are stored in clause_hash_table. Hash codes are calculated identically in the two programs. HERBY does not use a hash table to store information on clauses, however, as does THEO. This array, clause_hash_table, is normally set to store 2**20 entries, approximately 2,000,000 entries, each requiring four 32-bit words. The array occupies 32 megabytes of memory. Each entry in clause_hash_table has a typedef as shown here:

```
typedef struct {
        int      h_X,h_Y;       ; 64-bit hashcode of literal
        int      h_in;          ; iteration number
        int      h_ply;}        ; must be at same ply
        hnode;
```

cl stores information that describes the proof found. Each row of the array represents one inference of the proof as a 4-tuple. This information specifies which two clauses were resolved together and which literals of the two clauses were unified to produce the inference. This is sufficient to construct the proof.

In addition, just as in HERBY, there is an array, rand_array, where random numbers are initially placed and used to assign hash codes to clauses, as described in Chapter 9.

## 12.6 Functions Related to clause_hash_table

Thirteen functions related to clause_hash_table are in the file hash.c. The function is_clause_in_ht(h1) checks to see if a clause is in the hash table. The function add_clause_to_ht(h2) adds a clause to clause_hash_table after calling contradiction(h3) which checks to see whether the complementary clause, necessarily a unit clause, was previously found. Twelve memory locations are probed for an open location to store information on the new clause. If none is found, the new clause replaces the one of the twelve just probed that has the most literals. Is_unit_in_ht(h4) is used to see if one of the literals of a clause has been previously found as a unit clause and stored in the hash table; if so, under certain conditions, the clause need not be considered further. Is_comp_in_ht(h5) performs hash table resolutions, as discussed in Section 9.12.2. test_alt_hashcodes(h10), which calls is_test_in_ht(h11), checks to see whether a variant, say P(a), of some clause, say P(x), is in the hash table.

Several functions in the file hash.c are concerned with calculating the hash code of a literal (calc_lit_hc(h6)), or a variant of a literal, (calc_alt_hashcodes(h9), calc_alt_hash(h12)), or the hash code of a clause (calc_clause_hashcode(h13)). normalize_literal(h7) renames variables of a clause in normalized order.

## 12.7 Functions Related to cl

As THEO puts together the proof of a theorem, it saves the lines found in cl. The functions that save the lines are in the file pass2.c. save_line(ps2) calls elim_base_clause_hc(ps3) and elim_same_cont_hc(ps4) to determine whether hash codes, which must be found, are those of base clauses or clauses on the line being saved. If the hash codes are those of base clauses or on lines being saved, it is unnecessary to repeat the search to find them. They are deleted by delete_found_hcs(ps12) and delete_alt_hcs(ps13). move_down_renumber(ps14) renumbers clauses in the lines being saved for consistency.

When Phase 2 ends, THEO enters Phase 3 and checks the proof by calling search_proof_line(ps20), which in turn calls infer_proof_line(ps19), which in turn calls bres_proof_line(ps16) and factor_proof_line(ps17). These functions check the proof to ensure no hash error occurred.

## Exercises for Chapter 12

12.1. Examine the function format_clause(m15). Add print statements to the function to confirm the representation of the clauses of CLAUSEREP.THM shown in Section 11.4 . Exactly how was the theorem represented?

12.2. The code in the function unify is quite complex. Compose several theorems consisting of two clauses — one axiom, and one negated conclusion — that resolve to yield the NULL clause, and see if THEO performs as it should.

12.3. [Project] THEO's main shortcoming, relative to some of other better automated theorem-proving systems, is its handling of the Equal predicate. See if you can improve this capability. You may want to read Chapter 13 before trying this.

# 13 The CADE ATP System Competitions and Other Theorem Provers

For half a decade, the Conference on Automated Deduction has hosted an annual competition among automated theorem-proving systems. The first competition was held in 1996 at CADE-13 at Rutgers University. In 2000, the competition will take place at Carnegie Mellon University. THEO, under its previous name of TGTP, participated in the 1997 and 1998 competitions. A parallel version of TGTP called OCTOPUS participated in the 1997 competition. HERBY participated in the 1998 competition.

At each competition, there have been approximately a dozen or so participants from all around the world. Theorems are randomly selected from the TPTP Problem Library, avoiding those that are too hard and those that are too easy. Each entrant is given five minutes to solve each of the 60 to 80 theorems selected. The number of theorems has varied from year to year. There are several different competitions, but the main one involves solving a wide assortment of theorems. All entrants run on identical hardware, although there is a special category for systems that run on multiprocessing systems and other nonstandard hardware. Geoff Sutcliffe and Christian Suttner have co-organized the events.

Results of the competitions can be found in papers in the *Journal of Automated Reasoning* and on the World Wide Web at: www.cs.jcu.edu.au/~tptp/CASC–17/, www.cs.jcu.edu.au/~tptp/CASC–16, www.cs.jcu.edu.au/~tptp/CASC–15/, www.cs.jcu.edu.au/~tptp/CASC–14/, and www.jessen.informatik.tu-muenchen.de/~tptp/CASC–13/. Along with the results of the competitions, you can find the source code of all the programs that participated and download it to your computer. The code is intended to run on UNIX-based machines.

The following two sections examine proofs found for theorems in the 1997 competition by two automated theorem-proving programs, Gandalf and Otter. Gandalf finished in first place in that competition. Otter is one of the best-known programs. The output of the two programs is provided for three theorems, GEO026–2, GEO041–2, and GEO059–2. Note that the names of predicates, functions, and constants have been replaced by meaningless names so that the programs could not use this semantic knowledge to their advantage.

## 13.1 Gandalf

GEO026–2

```
Gandalf v. c-1.0b
starting to prove: /home/geoff/AtSUN/Problems/NEQRmEq_stfp/
otter/CCC001-1.p

clauses in usable:
-p1(x,y,z,z) | equal(x,y).
-p2(c5,c6,c7).
p2(x,f4(y,z,u,x,v),y) | -p2(v,x,u) | -p2(y,z,u).
p1(x,y,y,x).
p2(x,y,f8(x,y,z,u,v)) | -p2(y,z,u) | -p2(x,z,v) |
        equal(x,z).
p2(x,y,f9(x,z,u,y,v)) | -p2(z,u,y) | -p2(x,u,v) |
        equal(x,u).
p2(f8(x,y,z,u,v),v,f9(x,y,z,u,v)) | -p2(x,z,v) |
        -p2(y,z,u) | equal(x,z).
p1(x,f3(y,x,z,u),z,u).
-p1(x,y,z,u) | -p1(x,y,v,w) | p1(v,w,z,u).
p2(x,y,f3(x,y,z,u)).
-p1(x,y,x,z) | -p1(u,y,u,z) | -p1(v,y,v,z) |
        p2(v,u,x) | p2(x,v,u) | p2(u,x,v) | equal(y,z).
-p1(x,y,z,u) | -p1(v,w,x1,x2) | -p1(x,v,z,x1) |
        -p1(v,y,x1,u) | p1(w,y,x2,u) | -p2(z,x1,x2) |
        -p2(x,v,w) | equal(x,v).
equal(x,x).
p2(x,f10(y,z,x,u,v,w),w) | -p1(y,v,y,w) | -p1(y,z,y,x) |
        -p2(z,u,v) | -p2(y,z,v).
-p2(c6,c7,c5).
-p2(c7,c5,c6).
-p2(x,y,x) | equal(x,y).
p2(x,f4(y,x,z,u,v),v) | -p2(v,u,z) | -p2(y,x,z).
p1(x,y,x,f10(x,z,u,y,v,w)) | -p1(x,v,x,w) | -p1(x,z,x,u) |
        -p2(x,z,v) | -p2(z,y,v).
```

```
clauses in sos:
-equal(c13,c14).
p2(c11,c12,c13).
-equal(c11,c12).
p2(c11,c12,c14).
p1(c12,c13,c12,c14).

strategies selected:
((hyper 6 #t) (binary-unit 15 #t) (binary 30 #t 2 15)
(binary 69 #t) (binary-unit45 #f) (binary-order 21 #f)
(binary-nameorder 24 #f 2 15)
(binary-nameorder 30 #f (hyper-order 100000 #f))

using equality, using hyperresolution, using sos strategy,
using dynamic demodulation
proof attempt stopped: time limit

using binary resolution, using sos strategy, using unit
strategy, using dynamic demodulation
proof attempt stopped: time limit

using binary resolution, using sos strategy, using dynamic
demodulation, clause length limited to 15, clause depth
limited to 2
proof attempt stopped: time limit

using binary resolution, using sos strategy, using dynamic
demodulation
proof attempt stopped: time limit

using binary resolution, using unit strategy, using dynamic
demodulation
proof attempt stopped: time limit

using binary resolution, using term-depth-order strategy,
using dynamic demodulation
proof attempt stopped: time limit

using binary resolution, using pred-name-order strategy,
using dynamic demodulation, clause length limited to 15,
clause depth limited to 2

**** EMPTY CLAUSE DERIVED ****

timercheckpoints:
 c(24,0,6,6730,4,461,8251,5,634,8253,1,634,8277,0,634,14546,3,14
```

```
63,15621,4,1761,17264,5,2135,17276,5,2138,17280,1,2138,17304,0,2138,21339,3,3852,
22674,4,4389,26792,5,5139,26815,5,5141,26817,1,5142,26841,0,5142,
37524,3,8986,39911,4,10318,52304,5,12044,52362,5,12051,52362,1,12051,52386,0,
12051,69981,3,14363,76428,4,15430,82933,5,16552,82938,5,16556,82938,1,16556,
82962,0,16557,83646,3,17611,85183,4,18133,88280,5,18658,88295,5,18659,88341,1,
18665,88365,0,18666,94227,3,19869,94849,4,20468,95654,30,20771)

82942 [] p1(x,y,y,x).
82962 [] p1(c12,c13,c12,c14).
83012 [binary_s,82942,binary_s,82962] p1(c12,c14,c13,c12).
83033 [binary_s,82942,binary_s,83012] p1(c13,c12,c14,c12).
83036 [binary_s,82942,binary_s,83033] p1(c14,c12,c12,c13).
83505 [binary_s,82942,binary_s,83036] p1(c12,c14,c12,c13).
88342 [] -p1(x,y,z,z) | equal(x,y).
88345 [] p1(x,y,y,x).
88349 [] p1(x,f3(y,x,z,u),z,u).
88350 [] -p1(x,y,z,u) | -p1(x,y,v,w) | p1(v,w,z,u).
88353 [] -p1(x,y,z,u) | -p1(v,w,x1,x2) | -p1(x,v,z,x1) |
       -p1(v,y,x1,u) | p1(w,y,x2,u) | -p2(z,x1,x2) |
       -p2(x,v,w) | equal(x,v).
88361 [] -equal(c13,c14).
88362 [] p2(c11,c12,c13).
88363 [] -equal(c11,c12).
88364 [] p2(c11,c12,c14).
88365 [] p1(c12,c13,c12,c14).

88371 [factor,88350,binary_s,88349] p1(x,y,x,y).
88377 [factor,88353,factor_s,factor_s,factor_s,factor_s,
       factor_s,factor_s,binary_s,88371,
       binary_s,88371] -p1(x,y,x,z) | -p1(u,y,u,z) |
       p1(v,y,v,z) | -p2(x,u,v) | equal(x,u).
88388 [binary,88361,88342.2] -p1(c13,c14,x,x).
88498 [binary,88363,88377.5] -p1(c12,x,c12,y) |
       -p1(c11,x,c11,y) | p1(z,x,z,y) | -p2(c11,c12,z).
93238 [binary,88388,88498.3,binary_s,83505,binary_s,88362]
       -p1(c11,c14,c11,c13).
93632 [binary,88350.3,93238] -p1(x,y,c11,c14) |
       -p1(x,y,c11,c13).
93635 [binary,88345,93632] -p1(c14,c11,c11,c13).
93647 [binary,88350.3,93635] -p1(x,y,c14,c11) |
       -p1(x,y,c11,c13).
93738 [binary,88345,93647.2] -p1(c13,c11,c14,c11).
95655 [binary,88362,88353.7,binary_s,88363,binary_s,
       88371,binary_s,88365,binary_s,88364,
       binary_s,88371,binary_s,88371,binary_s,93738]
       contradiction.
```

```
statistics: --------------------------
given clauses:         3873       derived clauses:        177266
kept clauses:         53452       kept mid-nuclei:         33066
kept new demods:        120       forw unit-subs:          51177
forw double-subs:      6881       forw overdouble-subs:     5744
backward subs:          241       fast unit cutoff:         7829
full unit cutoff:        71       dbl  unit cutoff:           30

real runtime      :  210.0        process. runtime:       209.33

specific non-discr-tree subsumption statistics:
tried:              2278578       length fails:           154813
strength fails:      777340       predlist fails:         375587
aux str. fails:      245209       by-lit fails:           121457
full subs tried:     560944       full subs fail:         551808

real      3:32.1    user      3:29.6    sys          1.2
```

---

GEO041–2

```
Running /home/geoff/AtSUN/Systems/Gandalf/gandalf /home/
geoff/AtSUN/Problems/NEQ/RmEq_stfp/otter/CCC010-1.p
Time limit is 300s

Gandalf v. c-1.0b
starting to prove: /home/geoff/AtSUN/Problems/NEQ/
RmEq_stfp/otter/CCC010-1.p

clauses in usable:
p1(x,f3(y,x,z,u),z,u).
-p1(x,y,x,z) | -p1(u,y,u,z) | -p1(v,y,v,z) | p2(u,v,x) |
       p2(x,u,v) | p2(v,x,u) | equal(y,z).
p1(x,y,y,x).
-p2(c6,c7,c5).
-p1(x,y,z,u) | -p1(x,y,v,w) | p1(v,w,z,u).
p2(f8(x,y,z,u,v),v,f9(x,y,z,u,v)) | -p2(x,z,v) |
       -p2(y,z,u) | equal(x,z).
-p2(x,y,x) | equal(x,y).
-p1(x,y,z,z) | equal(x,y).
p1(x,y,x,f10(x,z,u,y,v,w)) | -p1(x,v,x,w) | -p1(x,z,x,u) |
       -p2(z,y,v) | -p2(x,z,v).
p2(x,y,f3(x,y,z,u)).
-p2(c7,c5,c6).
-p1(x,y,z,u) | -p1(v,w,x1,x2) | -p1(x,v,z,x1) |
       -p1(v,y,x1,u) | p1(w,y,x2,u) |
       -p2(x,v,w) | -p2(z,x1,x2) | equal(x,v).
```

```
p2(x,y,f9(x,z,u,y,v)) | -p2(z,u,y) | -p2(x,u,v) |
       equal(x,u).
p2(x,f4(y,z,u,x,v),y) | -p2(y,z,u) | -p2(v,x,u).
-p2(c5,c6,c7).
p2(x,y,f8(x,y,z,u,v)) | -p2(y,z,u) | -p2(x,z,v) |
       equal(x,z).
p2(x,f4(y,x,z,u,v),v) | -p2(y,x,z) | -p2(v,u,z).
equal(x,x).
p2(x,f10(y,z,x,u,v,w),w) | -p1(y,v,y,w) | -p1(y,z,y,x) |
       -p2(z,u,v) | -p2(y,z,v).
clauses in sos:
-equal(c12,c13).
p2(c11,c13,c12).
p2(c11,c12,c13).

strategies selected:
((hyper 6 #t) (binary-unit 15 #t) (binary 30 #t 2 15)
(binary 69 #t) (binary-unit 45 #f) (binary-order 21 #f)
(binary-nameorder 24 #f 2 15) (binary-nameorder 30 #f)
(hyper-order 100000 #f))

using equality, using hyperresolution, using sos strategy,
using dynamic demodulation
proof attempt stopped: time limit

using binary resolution, using sos strategy, using unit
strategy, using dynamic demodulation
proof attempt stopped: time limit

using binary resolution, using sos strategy, using dynamic
demodulation, clause length limited to 15, clause depth
limited to 2
proof attempt stopped: time limit

using binary resolution, using sos strategy, using dynamic
demodulation
proof attempt stopped: time limit

using binary resolution, using unit strategy, using dynamic
demodulation

********* EMPTY CLAUSE DERIVED *********

timer checkpoints:
c(22,0,6,5466,4,461,6955,5,607,6957,1,607,6979,0,607,13959,3,13
61,16217,4,1734,17604,5,2149,17738,5,2160,17743,1,2160,17765,0,2161,21515,3,3665,
23612,4,4414,27224,5,5162,27231,1,5226,27253,0,5226,37697,3,8677,41099,4,10403,4952
```

```
6,5,12127,49530,5,12129,49530,1,12129,49552,0,12130,52785,30,12344)

49531 [] p1(x,f3(y,x,z,u),z,u).
49537 [] -p2(x,y,x) | equal(x,y).
49538 [] -p1(x,y,z,z) | equal(x,y).
49540 [] p2(x,y,f3(x,y,z,u)).
49544 [] p2(x,f4(y,z,u,x,v),y) | -p2(y,z,u) | -p2(v,x,u).
49547 [] p2(x,f4(y,x,z,u,v),v) | -p2(y,x,z) | -p2(v,u,z).
49548 [] equal(x,x).
49550 [] -equal(c12,c13).
49551 [] p2(c11,c13,c12).
49552 [] p2(c11,c12,c13).
49583 [binary,49531,49538] equal(x,f3(y,x,z,z)).
49593 [para,49583.1.1.#f.(3),49540] p2(x,y,y).
49740 [binary,49593,49544.2] p2(x,f4(y,z,z,x,u),y) |
      -p2(u,x,z).
49796 [binary,49551,49740.2]
      p2(c13,f4(x,c12,c12,c13,c11),x).
49797 [binary,49552,49740.2]
      p2(c12,f4(x,c13,c13,c12,c11),x).
49898 [binary,49537,49796]
      equal(c13,f4(c13,c12,c12,c13,c11)).
49917 [para,49898.1.1.#f.(2),49547,binary_s,49593,
      binary_s, 49551] p2(c12,c13,c11).
49926 [binary,49544.2,49917]
      p2(x,f4(c12,c13,c11,x,y),c12) | -p2(y,x,c11).
49936 [binary,49537,49797]
      equal(c12,f4(c12,c13,c13,c12,c11)).
49955 [para,49936.1.1.#f.(2),49547,binary_s,49593,
      binary_s,49552] p2(c13,c12,c11).
52672 [binary,49955,49926.2]
      p2(c12,f4(c12,c13,c11,c12,c13),c12).
52756 [binary,49537,52672]
      equal(c12,f4(c12,c13,c11,c12,c13)).
52757 [para,52756.1.1.#f.(2),49547,binary_s,49917,
      binary_s,49955] p2(c13,c12,c13).
52758 [binary,49537,52757] equal(c13,c12).
52786 [para,52758.1.1.#t.(2),49550,binary_s,49548]
      contradiction.

statistics: ---------------------------
given clauses:         2254    derived clauses:         93828
kept clauses:         33691    kept mid-nuclei:         16597
kept new demods:        105    forw unit-subs:          19894
forw double-subs:      9066    forw overdouble-subs:     3952
backward subs:          502    fast unit cutoff:         2382
```

```
full unit cutoff:      68       dbl  unit cutoff:          7

real runtime     :  123.0        process. runtime:     123.77

specific non-discr-tree subsumption statistics:
tried:              2047695      length fails:         195174
strength fails:      697634      predlist fails:       210469
aux str. fails:      297239      by-lit fails:         213282
full subs tried:     370393      full subs fail:       363469

real      2:05.3          user      2:04.1          sys       1.0

-----------------------------------------------------------
```

GEO059–2

```
Running /home/geoff/AtSUN/Systems/Gandalf/gandalf /home/
geoff/AtSUN/Problems/NEQ/RmEq_stfp/otter/CCC020-1.p
Time limit is 300s

Gandalf v. c-1.0b
starting to prove: /home/geoff/AtSUN/Problems/NEQ/
RmEq_stfp/otter/CCC020-1.p

clauses in usable:
p2(x,y,f3(x,y,z,u)).
p2(x,f10(y,z,x,u,v,w),w) | -p1(y,z,y,x) | -p1(y,v,y,w) |
      -p2(z,u,v) | -p2(y,z,v).
p1(x,f3(y,x,z,u),z,u).
equal(f11(x,y),f3(x,y,x,y)).
-p1(x,y,z,u) | -p1(x,y,v,w) | p1(z,u,v,w).
p1(x,y,y,x).
p2(f8(x,y,z,u,v),v,f9(x,y,z,u,v)) | -p2(x,z,v) |
      -p2(y,z,u) | equal(x,z).
-p1(x,y,x,z) | -p1(u,y,u,z) | -p1(v,y,v,z) |
      p2(x,u,v) | p2(u,v,x) | p2(v,x,u) | equal(y,z).
-p2(x,y,x) | equal(x,y).
-p1(x,y,z,z) | equal(x,y).
-p2(c7,c5,c6).
p2(x,y,f8(x,y,z,u,v)) | -p2(x,z,v) | -p2(y,z,u) |
      equal(x,z).
p1(x,y,x,f10(x,z,u,y,v,w)) | -p1(x,z,x,u) | -p1(x,v,x,w) |
      -p2(z,y,v) | -p2(x,z,v).
-p2(c6,c7,c5).
p2(x,f4(y,x,z,u,v),v) | -p2(v,u,z) | -p2(y,x,z).
p2(x,f4(y,z,u,x,v),y) | -p2(y,z,u) | -p2(v,x,u).
-p2(c5,c6,c7).
```

```
-p1(x,y,z,u) | -p1(x,v,z,w) | -p1(x1,y,x2,u) |
       -p1(x1,x,x2,z) | p1(v,y,w,u) | -p2(x2,z,w) |
       -p2(x1,x,v) | equal(x1,x).
p2(x,y,f9(x,z,u,y,v)) | -p2(x,u,v) | -p2(z,u,y) |
       equal(x,u).
equal(x,x).

clauses in sos:
-p1(c12,c13,c12,f11(f11(c13,c12),c12)).

strategies selected:
((hyper 6 #t) (binary-unit 15 #t) (binary 30 #t 3 15)
(binary 69 #t) (binary-unit 45 #f) (binary-order 21 #f)
(binary-nameorder 24 #f 3 15) (binary-nameorder 30 #f)
(hyper-order 100000 #f))

using equality, using hyperresolution, using sos strategy,
using dynamic demodulation
proof attempt stopped: sos exhausted

using binary resolution, using sos strategy, using unit
strategy, using dynamic demodulation

**** EMPTY CLAUSE DERIVED ****

timer checkpoints: c(21,0,6,42,0,7,1309,30,145)

24 [] p1(x,f3(y,x,z,u),z,u).
25 [] equal(f11(x,y),f3(x,y,x,y)).
26 [] -p1(x,y,z,u) | -p1(x,y,v,w) | p1(z,u,v,w).
27 [] p1(x,y,y,x).
42 [] -p1(c12,c13,c12,f11(f11(c13,c12),c12)).
45 [binary,26.3,42] -p1(x,y,c12,f11(f11(c13,c12),c12)) |
       -p1(x,y,c12,c13).
57 [binary,27,45.2] -p1(c13,c12,c12,f11(f11(c13,c12),c12)).
74 [binary,26.3,57] -p1(x,y,c12,f11(f11(c13,c12),c12)) |
       -p1(x,y,c13,c12).
165 [binary,24,74.2]
       -p1(x,f3(y,x,c13,c12),c12,f11(f11(c13,c12),c12)).
179 [para,25.1.1.#f.(2),165]
       -p1(c12,f11(c13,c12),c12,f11(f11(c13,c12),c12)).
186 [binary,26.3,179] -p1(x,y,c12,f11(f11(c13,c12),c12)) |
       -p1(x,y,c12,f11(c13,c12)).
715 [binary,27,186.2]
       -p1(f11(c13,c12),c12,c12,f11(f11(c13,c12),c12)).
724 [binary,26.3,715] -p1(x,y,c12,f11(f11(c13,c12),c12)) |
       -p1(x,y,f11(c13,c12),c12).
```

```
1265 [binary,24,724.2]
      -p1(x,f3(y,x,f11(c13,c12),c12),c12,f11
            (f11(c13,c12),c12)).
1298 [para,25.1.1.#f.(2),1265]
      -p1(c12,f11(f11(c13,c12),c12),c12,f11(f11(c13,c12),
            c12)).
1310 [binary,26.3,1298,binary_s,27] contradiction.

statistics: --------------------------
given clauses:          254         derived clauses:        1844
kept clauses:          1265         kept mid-nuclei:           0
kept new demods:          0         forw unit-subs:          357
forw double-subs:       137         forw overdouble-subs:     84
backward subs:            3         fast unit cutoff:          0
full unit cutoff:         0         dbl  unit cutoff:          0

real runtime     :      1.0         process. runtime:       1.46

specific non-discr-tree subsumption statistics:
tried:                10230         length fails:           1091
strength fails:         178         predlist fails:         2425
aux str. fails:         465         by-lit fails:              6
full subs tried:       4282         full subs fail:         4218


real          1.7         user          1.5         sys           0.0
```

## 13.2 Otter

GEO026–2

```
Running /home/geoff/AtSUN/Systems/Otter/otter /home/geoff/
AtSUN/Problems/NEQ/RmEq_stfp/otter/CCC001-1.p
Time limit is 300s

----- Otter 3.0.5-beta, July 1997 -----

The job was started by ??? on ???, Wed Jul 16 10:40:47 1997
The command was "/home/geoff/AtSUN/Systems/Otter/otter305-
beta".

set(prolog_style_variables).
set(auto).
   dependent: set(auto2).
```

```
   dependent: set(process_input).
   dependent: clear(print_kept).
   dependent: clear(print_new_demod).
   dependent: clear(print_back_demod).
   dependent: clear(print_back_sub).
   dependent: set(control_memory).
   dependent: assign(max_mem, 20000).
   dependent: assign(pick_given_ratio, 4).
   dependent: assign(stats_level, 1).
dependent: assign(max_seconds, 10800).
set(tptp_eq).
clear(print_given).
assign(max_seconds,300).

list(usable).
0 [] equal(X1,X2) | -p1(X1,X2,X3,X3).
0 [] -p2(c5,c6,c7).
0 [] p2(X4,f4(X1,X2,X3,X4,X5),X1) | -p2(X5,X4,X3) |
      -p2(X1,X2,X3).
0 [] p1(X1,X2,X2,X1).
0 [] -p2(X1,X3,X5) | equal(X1,X3) | -p2(X2,X3,X4) |
      p2(X1,X2,f8(X1,X2,X3,X4,X5)).
0 [] -p2(X1,X3,X5) | equal(X1,X3) |
      p2(X1,X4,f9(X1,X2,X3,X4,X5)) | -p2(X2,X3,X4).
0 [] equal(X1,X3) | -p2(X2,X3,X4) | -p2(X1,X3,X5) |
      p2(f8(X1,X2,X3,X4,X5),X5,f9(X1,X2,X3,X4,X5)).
0 [] p1(X4,f3(X3,X4,X2,X1),X2,X1).
0 [] p1(X5,X1,X6,X2) | -p1(X3,X4,X6,X2) | -p1(X3,X4,X5,X1).
0 [] p2(X3,X4,f3(X3,X4,X2,X1)).
0 [] -p1(X4,X2,X4,X1)|p2(X5,X3,X4) | equal(X2,X1) |
      -p1(X3,X2,X3,X1) | -p1(X5,X2,X5,X1) |p2(X4,X5,X3) |
      p2(X3,X4,X5).
0 [] -p1(X3,X1,X7,X5) | equal(X2,X3) | -p2(X2,X3,X4) |
      -p1(X2,X3,X6,X7) | p1(X4,X1,X8,X5) |
      -p1(X3,X4,X7,X8) | -p2(X6,X7,X8) | -p1(X2,X1,X6,X5).
0 [] equal(X1,X1).
0 [] -p2(X1,X2,X4) | -p1(X1,X2,X1,X5) | -p1(X1,X4,X1,X6) |
      -p2(X2,X3,X4) | p2(X5,f10(X1,X2,X5,X3,X4,X6),X6).
0 [] -p2(c6,c7,c5).
0 [] -p2(c7,c5,c6).
0 [] equal(X1,X2) | -p2(X1,X2,X1).
0 [] -p2(X1,X2,X3) | p2(X2,f4(X1,X2,X3,X4,X5),X5) |
      -p2(X5,X4,X3).
0 [] -p1(X1,X2,X1,X5) | -p2(X2,X3,X4) |
      p1(X1,X3,X1,f10(X1,X2,X5,X3,X4,X6)) | -p2(X1,X2,X4) |
      -p1(X1,X4,X1,X6).
end_of_list.
```

```
list(sos).
0 [] -equal(c13,c14).
0 [] p2(c11,c12,c13).
0 [] -equal(c11,c12).
0 [] p2(c11,c12,c14).
0 [] p1(c12,c13,c12,c14).
end_of_list.

Every positive clause in sos is ground (or sos is empty);
therefore we move all positive usable clauses to sos.

Properties of input clauses: prop=0, horn=0, equality=1,
symmetry=0, max_lits=8.

Setting hyper_res, because there are nonunits.
   dependent: set(hyper_res).
Setting ur_res, because this is a nonunit set containing
either equality literals or non-Horn clauses.
   dependent: set(ur_res).
Setting factor and  unit_deletion, because there are non-
Horn clauses.
   dependent: set(factor).
   dependent: set(unit_deletion).
Equality is present, so we set the knuth_bendix flag.
   dependent: set(knuth_bendix).
   dependent: set(para_from).
   dependent: set(para_into).
   dependent: clear(para_from_right).
   dependent: clear(para_into_right).
   dependent: set(para_from_vars).
   dependent: set(eq_units_both_ways).
   dependent: set(dynamic_demod_all).
   dependent: set(dynamic_demod).
   dependent: set(order_eq).
   dependent: set(back_demod).
   dependent: set(lrpo).
As an incomplete heuristic, we paramodulate with units
only.
   dependent: set(para_from_units_only).
   dependent: set(para_into_units_only).


------------> process usable:
** KEPT (pick-wt=8): 1 [] equal(A,B) | -p1(A,B,C,C).
** KEPT (pick-wt=4): 2 [] -p2(c5,c6,c7).
** KEPT (pick-wt=17): 3 [] p2(A,f4(B,C,D,A,E),B) |
      -p2(E,A,D) | -p2(B,C,D).
```

```
** KEPT (pick-wt=20): 4 [] -p2(A,B,C)|equal(A,B) |
      -p2(D,B,E) | p2(A,D,f8(A,D,B,E,C)).
** KEPT (pick-wt=20): 5 [] -p2(A,B,C) | equal(A,B) |
      p2(A,D,f9(A,E,B,D,C)) | -p2(E,B,D).
** KEPT (pick-wt=25): 6 [] equal(A,B) | -p2(C,B,D) |
      -p2(A,B,E) | p2(f8(A,C,B,D,E),E,f9(A,C,B,D,E)).
** KEPT (pick-wt=15): 7 [] p1(A,B,C,D) | -p1(E,F,C,D) |
      -p1(E,F,A,B).
** KEPT (pick-wt=30): 8 [] -p1(A,B,A,C) | p2(D,E,A) |
      equal(B,C) | -p1(E,B,E,C)| -p1(D,B,D,C) |
      p2(A,D,E) | p2(E,A,D).
** KEPT (pick-wt=36): 9 [] -p1(A,B,C,D) | equal(E,A) |
      -p2(E,A,F) | -p1(E,A,G,C) | p1(F,B,H,D) |
      -p1(A,F,C,H) | -p2(G,C,H) | -p1(E,B,G,D).
** KEPT (pick-wt=28): 10 [] -p2(A,B,C) | -p1(A,B,A,D) |
      -p1(A,C,A,E) | p2(B,F,C) | p2(D,f10(A,B,D,F,C,E),E).
** KEPT (pick-wt=4): 11 [] -p2(c6,c7,c5).
** KEPT (pick-wt=4): 12 [] -p2(c7,c5,c6).
** KEPT (pick-wt=7): 13 [] equal(A,B)| -p2(A,B,A).
** KEPT (pick-wt=17): 14 [] -p2(A,B,C) |
      p2(B,f4(A,B,C,D,E),E) | -p2(E,D,C).
** KEPT (pick-wt=29): 15 [] -p1(A,B,A,C) | -p2(B,D,E) |
      p1(A,D,A,f10(A,B,C,D,E,F)) | -p2(A,B,E) |
      -p1(A,E,A,F).
------------> process sos:
** KEPT (pick-wt=3): 35 [copy,34,flip.1] -equal(c14,c13).
** KEPT (pick-wt=4): 36 [] p2(c11,c12,c13).
** KEPT (pick-wt=3): 38 [copy,37,flip.1] -equal(c12,c11).
** KEPT (pick-wt=4): 39 [] p2(c11,c12,c14).
** KEPT (pick-wt=5): 40 [] p1(c12,c13,c12,c14).
** KEPT (pick-wt=5): 41 [] p1(A,B,B,A).
** KEPT (pick-wt=9): 42 [] p1(A,f3(B,A,C,D),C,D).
** KEPT (pick-wt=8): 43 [] p2(A,B,f3(A,B,C,D)).
** KEPT (pick-wt=3): 44 [] equal(A,A).

41 back subsumes 30.

Following clause subsumed by 44 during input processing:
0 [copy,44,flip.1] equal(A,A).

======= end of input processing =======
=========== start of search ===========


-------- PROOF ------------>
UNIT CONFLICT at   0.26 sec
----> 90 [binary,89.1,84.1] $F.
```

```
Length of proof is 8.  Level of proof is 3.
---------------- PROOF ----------------
1 [] equal(A,B) | -p1(A,B,C,C).
2 [] -p2(c5,c6,c7).
3 [] p2(A,f4(B,C,D,A,E),B) | -p2(E,A,D) | -p2(B,C,D).
7 [] p1(A,B,C,D) | -p1(E,F,C,D) | -p1(E,F,A,B).
9 [] -p1(A,B,C,D) | equal(E,A) | -p2(E,A,F) |
       -p1(E,A,G,C) | p1(F,B,H,D) | -p1(A,F,C,H)|
       -p2(G,C,H) | -p1(E,B,G,D).
20 [factor,7,2,3] p1(A,B,A,B) | -p1(C,D,A,B).
34 [] -equal(c13,c14).
35 [copy,34,flip.1] -equal(c14,c13).
36 [] p2(c11,c12,c13).
37 [] -equal(c11,c12).
38 [copy,37,flip.1] -equal(c12,c11).
39 [] p2(c11,c12,c14).
40 [] p1(c12,c13,c12,c14).
41 [] p1(A,B,B,A).
42 [] p1(A,f3(B,A,C,D),C,D).
46 [ur,35,1] -p1(c14,c13,A,A).
74 [hyper,41,20] p1(A,B,A,B).
81,80 [hyper,42,1,flip.1] equal(f3(A,B,C,C),B).
84 [ur,46,7,42,demod,81] -p1(A,A,c14,c13).
89 [hyper,74,9,74,36,74,40,39,unit_del,38] p1(c13,A,c14,A).
90 [binary,89.1,84.1] $F.

------------ end of proof -------------

Search stopped by max_proofs option.
============ end of search ============

-------------- statistics -------------
clauses given                 14
clauses generated            369
clauses kept                  86
clauses forward subsumed     298
clauses back subsumed          4
Kbytes malloced              191

----------- times (seconds) -----------
user CPU time          0.26          (0 hr, 0 min, 0 sec)
system CPU time        0.00          (0 hr, 0 min, 0 sec)
wall-clock time        1             (0 hr, 0 min, 1 sec)
hyper_res time         0.00
UR_res time            0.00
para_into time         0.00
para_from time         0.00
```

```
for_sub time            0.00
back_sub time           0.00
conflict time           0.00
demod time              0.00

The job finished         Wed Jul 16 10:40:48 1997

real          0.4         user          0.3           sys          0.0
```

---

GEO041-2

```
Running /home/geoff/AtSUN/Systems/Otter/otter /home/geoff/
AtSUN/Problems/NEQ/RmEq_stfp/otter/CCC010-1.p
Time limit is 300s

----- Otter 3.0.5-beta, July 1997 -----

The job was started by ??? on ???, Wed Jul 16 12:11:12 1997
The command was "/home/geoff/AtSUN/Systems/Otter/otter305-
beta".

set(prolog_style_variables).
set(auto).
   dependent: set(auto2).
   dependent: set(process_input).
   dependent: clear(print_kept).
   dependent: clear(print_new_demod).
   dependent: clear(print_back_demod).
   dependent: clear(print_back_sub).
   dependent: set(control_memory).
   dependent: assign(max_mem, 20000).
   dependent: assign(pick_given_ratio, 4).
   dependent: assign(stats_level, 1).
   dependent: assign(max_seconds, 10800).
set(tptp_eq).
clear(print_given).
assign(max_seconds,300).

list(usable).
0 [] p1(X4,f3(X3,X4,X2,X1),X2,X1).
0 [] -p1(X5,X2,X5,X1) | p2(X3,X4,X5) | p2(X5,X3,X4) |
      p2(X4,X5,X3) | equal(X2,X1) | -p1(X3,X2,X3,X1) |
      -p1(X4,X2,X4,X1).
0 [] p1(X1,X2,X2,X1).
0 [] -p2(c6,c7,c5).
0 [] -p1(X3,X4,X5,X1) | -p1(X3,X4,X6,X2) | p1(X5,X1,X6,X2).
```

```
0 [] equal(X1,X3) |
      p2(f8(X1,X2,X3,X4,X5),X5,f9(X1,X2,X3,X4,X5)) |
      -p2(X1,X3,X5) | -p2(X2,X3,X4).
0 [] equal(X1,X2) | -p2(X1,X2,X1).
0 [] equal(X1,X2) | -p1(X1,X2,X3,X3).
0 [] p1(X1,X3,X1,f10(X1,X2,X5,X3,X4,X6)) | -p2(X1,X2,X4) |
      -p1(X1,X2,X1,X5) | -p1(X1,X4,X1,X6) | -p2(X2,X3,X4).
0 [] p2(X3,X4,f3(X3,X4,X2,X1)).
0 [] -p2(c7,c5,c6).
0 [] -p2(X2,X3,X4) | -p2(X6,X7,X8)| -p1(X3,X1,X7,X5) |
      p1(X4,X1,X8,X5) | -p1(X2,X3,X6,X7) | equal(X2,X3)|
      -p1(X2,X1,X6,X5) | -p1(X3,X4,X7,X8).
0 [] p2(X1,X4,f9(X1,X2,X3,X4,X5)) | -p2(X1,X3,X5)|
      -p2(X2,X3,X4) | equal(X1,X3).
0 [] -p2(X5,X4,X3) | p2(X4,f4(X1,X2,X3,X4,X5),X1) |
      -p2(X1,X2,X3).
0 [] -p2(c5,c6,c7).
0 [] equal(X1,X3) | -p2(X1,X3,X5) |
      -p2(X2,X3,X4) | p2(X1,X2,f8(X1,X2,X3,X4,X5)).
0 [] p2(X2,f4(X1,X2,X3,X4,X5),X5) | -p2(X1,X2,X3) |
      -p2(X5,X4,X3).
0 [] equal(X1,X1).
0 [] -p2(X2,X3,X4)| -p2(X1,X2,X4) |
      p2(X5,f10(X1,X2,X5,X3,X4,X6),X6) |
      -p1(X1,X4,X1,X6) | -p1(X1,X2,X1,X5).
end_of_list.

list(sos).
0 [] -equal(c12,c13).
0 [] p2(c11,c13,c12).
0 [] p2(c11,c12,c13).
end_of_list.

Every positive clause in sos is ground (or sos is empty);
therefore we move all positive usable clauses to sos.

Properties of input clauses: prop=0, horn=0, equality=1,
symmetry=0, max_lits=8.

Setting hyper_res, because there are nonunits.
   dependent: set(hyper_res).
Setting ur_res, because this is a nonunit set containing
either equality literals or non-Horn clauses.
   dependent: set(ur_res).
Setting factor and  unit_deletion, because there are non-
Horn clauses.
   dependent: set(factor).
```

```
   dependent: set(unit_deletion).
Equality is present, so we set the knuth_bendix flag.
   dependent: set(knuth_bendix).
   dependent: set(para_from).
   dependent: set(para_into).
   dependent: clear(para_from_right).
   dependent: clear(para_into_right).
   dependent: set(para_from_vars).
   dependent: set(eq_units_both_ways).
   dependent: set(dynamic_demod_all).
   dependent: set(dynamic_demod).
   dependent: set(order_eq).
   dependent: set(back_demod).
   dependent: set(lrpo).
As an incomplete heuristic, we paramodulate with units only.
   dependent: set(para_from_units_only).
   dependent: set(para_into_units_only).


------------> process usable:
** KEPT (pick-wt=30): 1 [] -p1(A,B,A,C) | p2(D,E,A) |
      p2(A,D,E) | p2(E,A,D) | equal(B,C) | -p1(D,B,D,C) |
      -p1(E,B,E,C).
** KEPT (pick-wt=4): 2 [] -p2(c6,c7,c5).
** KEPT (pick-wt=15): 3 [] -p1(A,B,C,D) | -p1(A,B,E,F) |
      p1(C,D,E,F).
** KEPT (pick-wt=25): 4 [] equal(A,B) |
      p2(f8(A,C,B,D,E),E,f9(A,C,B,D,E)) | -p2(A,B,E)|
      -p2(C,B,D).
** KEPT (pick-wt=7): 5 [] equal(A,B) | -p2(A,B,A).
** KEPT (pick-wt=8): 6 [] equal(A,B) | -p1(A,B,C,C).
** KEPT (pick-wt=29): 7 [] p1(A,B,A,f10(A,C,D,B,E,F)) |
      -p2(A,C,E) | -p1(A,C,A,D) | -p1(A,E,A,F) |
      -p2(C,B,E).
** KEPT (pick-wt=4): 8 [] -p2(c7,c5,c6).
** KEPT (pick-wt=36): 9 [] -p2(A,B,C) | -p2(D,E,F)|
      -p1(B,G,E,H) | p1(C,G,F,H) | -p1(A,B,D,E) |
equal(A,B) | -p1(A,G,D,H) | -p1(B,C,E,F).
** KEPT (pick-wt=20): 10 [] p2(A,B,f9(A,C,D,B,E)) |
      -p2(A,D,E)| -p2(C,D,B)| equal(A,D).
** KEPT (pick-wt=17): 11 [] -p2(A,B,C) |
p2(B,f4(D,E,C,B,A),D) | -p2(D,E,C).
** KEPT (pick-wt=4): 12 [] -p2(c5,c6,c7).
** KEPT (pick-wt=20): 13 [] equal(A,B) | -p2(A,B,C) |
      -p2(D,B,E) | p2(A,D,f8(A,D,B,E,C)).
** KEPT (pick-wt=17): 14 [] p2(A,f4(B,A,C,D,E),E) |
      -p2(B,A,C) | -p2(E,D,C).
```

```
** KEPT (pick-wt=28): 15 [] -p2(A,B,C) | -p2(D,A,C) |
      p2(E,f10(D,A,E,B,C,F),F) | -p1(D,C,D,F) |
      -p1(D,A,D,E).

------------> process sos:
** KEPT (pick-wt=3): 35 [copy,34,flip.1] -equal(c13,c12).
** KEPT (pick-wt=4): 36 [] p2(c11,c13,c12).
** KEPT (pick-wt=4): 37 [] p2(c11,c12,c13).
** KEPT (pick-wt=9): 38 [] p1(A,f3(B,A,C,D),C,D).
** KEPT (pick-wt=5): 39 [] p1(A,B,B,A).
** KEPT (pick-wt=8): 40 [] p2(A,B,f3(A,B,C,D)).
** KEPT (pick-wt=3): 41 [] equal(A,A).
39 back subsumes 31.

Following clause subsumed by 41 during input processing:
0 [copy,41,flip.1] equal(A,A).

======= end of input processing =======
=========== start of search ===========

Resetting weight limit to 14.
Resetting weight limit to 14.
sos_size=5385

-- HEY ???, WE HAVE A PROOF!! --

--> UNIT CONFLICT at  87.29 sec
--> 6266 [binary, 6265.1,43.1] $F.

Length of proof is 16.  Level of proof is 8.
---------------- PROOF ----------------

5 [] equal(A,B) | -p2(A,B,A).
6 [] equal(A,B) | -p1(A,B,C,C).
11 [] -p2(A,B,C) | p2(B,f4(D,E,C,B,A),D) | -p2(D,E,C).
14 [] p2(A,f4(B,A,C,D,E),E) | -p2(B,A,C) | -p2(E,D,C).
34 [] -equal(c12,c13).
35 [copy,34,flip.1] -equal(c13,c12).
36 [] p2(c11,c13,c12).
37 [] p2(c11,c12,c13).
38 [] p1(A,f3(B,A,C,D),C,D).
40 [] p2(A,B,f3(A,B,C,D)).
43 [ur,35,5] -p2(c13,c12,c13).
54 [hyper,38,6,flip.1] equal(f3(A,B,C,C),B).
104 [para_from,54.1.1,40.1.3] p2(A,B,B).
128 [hyper,104,14,37] p2(c13,f4(A,c13,c13,c12,c11),c11).
129 [hyper,104,14,36] p2(c12,f4(A,c12,c12,c13,c11),c11).
```

```
139 [hyper,104,11,37] p2(c12,f4(A,c13,c13,c12,c11),A).
140 [hyper,104,11,36] p2(c13,f4(A,c12,c12,c13,c11),A).
1589 [hyper,139,5,flip.1]
      equal(f4(c12,c13,c13,c12,c11),c12).
1598 [para_from,1589.1.1,128.1.2] p2(c13,c12,c11).
1928 [hyper,140,5,flip.1]
      equal(f4(c13,c12,c12,c13,c11),c13).
1937 [para_from,1928.1.1,129.1.2] p2(c12,c13,c11).
2329 [hyper,1937,14,1598]
      p2(c12,f4(c13,c12,c11,c13,c12),c12).
2378 [hyper,1937,11,1598]
      p2(c13,f4(c13,c12,c11,c13,c12),c13).
6264,6263 [hyper,2329,5,flip.1]
      equal(f4(c13,c12,c11,c13,c12),c12).
6265 [back_demod,2378,demod,6264] p2(c13,c12,c13).
6266 [binary,6265.1,43.1] $F.

------------ end of proof -------------

Search stopped by max_proofs option.

============ end of search ============

-------------- statistics -------------
clauses given                  133
clauses generated           105351
clauses kept                  6237
clauses forward subsumed     31894
clauses back subsumed          345
Kbytes malloced               6993

----------- times (seconds) -----------
user CPU time         87.29          (0 hr, 1 min, 27 sec)
system CPU time        0.17          (0 hr, 0 min, 0 sec)
wall-clock time          88          (0 hr, 1 min, 28 sec)
hyper_res time         0.00
UR_res time            0.00
para_into time         0.00
para_from time         0.00
for_sub time           0.00
back_sub time          0.00
conflict time          0.00
demod time             0.00

The job finished        Wed Jul 16 12:12:40 1997

real     1:27.7      user     1:27.4      sys        0.2
```

GEO059–2

```
Running /home/geoff/AtSUN/Systems/Otter/otter /home/geoff/
AtSUN/Problems/NEQ/RmEq_stfp/otter/CCC020-1.p
Time limit is 300s

----- Otter 3.0.5-beta, July 1997 -----

The job was started by ??? on ???, Wed Jul 16 14:06:58 1997
The command was "/home/geoff/AtSUN/Systems/Otter/otter305-
beta".

set(prolog_style_variables).
set(auto).
   dependent: set(auto2).
   dependent: set(process_input).
   dependent: clear(print_kept).
   dependent: clear(print_new_demod).
   dependent: clear(print_back_demod).
   dependent: clear(print_back_sub).
   dependent: set(control_memory).
   dependent: assign(max_mem, 20000).
   dependent: assign(pick_given_ratio, 4).
   dependent: assign(stats_level, 1).
   dependent: assign(max_seconds, 10800).
set(tptp_eq).
clear(print_given).
assign(max_seconds,300).
list(usable).
0 [] p2(X3,X4,f3(X3,X4,X2,X1)).
0 [] p2(X5,f10(X1,X2,X5,X3,X4,X6),X6) | -p1(X1,X4,X1,X6) |
       -p1(X1,X2,X1,X5) | -p2(X2,X3,X4) | -p2(X1,X2,X4).
0 [] p1(X4,f3(X3,X4,X2,X1),X2,X1).
0 [] equal(f11(X1,X2),f3(X1,X2,X1,X2)).
0 [] -p1(X3,X4,X6,X2) | p1(X5,X1,X6,X2) | -p1(X3,X4,X5,X1).
0 [] p1(X1,X2,X2,X1).
0 [] -p2(X2,X3,X4) | equal(X1,X3) | -p2(X1,X3,X5) |
       p2(f8(X1,X2,X3,X4,X5),X5,f9(X1,X2,X3,X4,X5)).
0 [] -p1(X5,X2,X5,X1) | p2(X5,X3,X4) | equal(X2,X1) |
       p2(X3,X4,X5) | p2(X4,X5,X3) | -p1(X3,X2,X3,X1) |
       -p1(X4,X2,X4,X1).
0 [] -p2(X1,X2,X1) | equal(X1,X2).
0 [] equal(X1,X2) | -p1(X1,X2,X3,X3).
0 [] -p2(c7,c5,c6).
0 [] equal(X1,X3) | -p2(X2,X3,X4) | -p2(X1,X3,X5) |
       p2(X1,X2,f8(X1,X2,X3,X4,X5)).
```

```
0 [] -p2(X2,X3,X4) | -p2(X1,X2,X4) | -p1(X1,X4,X1,X6) |
       p1(X1,X3,X1,f10(X1,X2,X5,X3,X4,X6)) |
       -p1(X1,X2,X1,X5).
0 [] -p2(c6,c7,c5).
0 [] p2(X2,f4(X1,X2,X3,X4,X5),X5) | -p2(X5,X4,X3) |
       -p2(X1,X2,X3).
0 [] -p2(X5,X4,X3) | p2(X4,f4(X1,X2,X3,X4,X5),X1) |
       -p2(X1,X2,X3).
0 [] -p2(c5,c6,c7).
0 [] p1(X4,X1,X8,X5) | -p1(X2,X3,X6,X7) | -p2(X2,X3,X4) |
       equal(X2,X3) | -p1(X2,X1,X6,X5) | -p2(X6,X7,X8) |
       -p1(X3,X1,X7,X5) | -p1(X3,X4,X7,X8).
0 [] -p2(X2,X3,X4) | equal(X1,X3) | -p2(X1,X3,X5) |
       p2(X1,X4,f9(X1,X2,X3,X4,X5)).
0 [] equal(X1,X1).
end_of_list.

list(sos).
0 [] -p1(c12,c13,c12,f11(f11(c13,c12),c12)).
end_of_list.

Every positive clause in sos is ground (or sos is empty);
therefore we move all positive usable clauses to sos.

Properties of input clauses: prop=0, horn=0, equality=1,
symmetry=0, max_lits=8.
Setting hyper_res, because there are nonunits.
   dependent: set(hyper_res).
Setting ur_res, because this is a nonunit set containing
either equality literals or non-Horn clauses.
   dependent: set(ur_res).
Setting factor and  unit_deletion, because there are non-
Horn clauses.
   dependent: set(factor).
   dependent: set(unit_deletion).
Equality is present, so we set the knuth_bendix flag.
   dependent: set(knuth_bendix).
   dependent: set(para_from).
   dependent: set(para_into).
   dependent: clear(para_from_right).
   dependent: clear(para_into_right).
   dependent: set(para_from_vars).
   dependent: set(eq_units_both_ways).
   dependent: set(dynamic_demod_all).
   dependent: set(dynamic_demod).
   dependent: set(order_eq).
```

```
   dependent: set(back_demod).
   dependent: set(lrpo).
As an incomplete heuristic, we paramodulate with units only.
   dependent: set(para_from_units_only).
   dependent: set(para_into_units_only).

------------> process usable:
** KEPT (pick-wt=28): 1 [] p2(A,f10(B,C,A,D,E,F),F) |
       -p1(B,E,B,F) | -p1(B,C,B,A) | -p2(C,D,E) |
       -p2(B,C,E).
** KEPT (pick-wt=15): 2 [] -p1(A,B,C,D) | p1(E,F,C,D) |
       -p1(A,B,E,F).
** KEPT (pick-wt=25): 3 [] -p2(A,B,C) | equal(D,B) |
       -p2(D,B,E) | p2(f8(D,A,B,C,E),E,f9(D,A,B,C,E)).
** KEPT (pick-wt=30): 4 [] -p1(A,B,A,C) | p2(A,D,E) |
       equal(B,C) | p2(D,E,A) | p2(E,A,D) | -p1(D,B,D,C) |
       -p1(E,B,E,C).
** KEPT (pick-wt=7): 5 [] -p2(A,B,A) | equal(A,B).
** KEPT (pick-wt=8): 6 [] equal(A,B) | -p1(A,B,C,C).
** KEPT (pick-wt=4): 7 [] -p2(c7,c5,c6).
** KEPT (pick-wt=20): 8 [] equal(A,B) | -p2(C,B,D) |
       -p2(A,B,E) | p2(A,C,f8(A,C,B,D,E)).
** KEPT (pick-wt=29): 9 [] -p2(A,B,C) | -p2(D,A,C) |
       -p1(D,C,D,E) | p1(D,B,D,f10(D,A,F,B,C,E)) |
       -p1(D,A,D,F).
** KEPT (pick-wt=4): 10 [] -p2(c6,c7,c5).
** KEPT (pick-wt=17): 11 [] p2(A,f4(B,A,C,D,E),E) |
       -p2(E,D,C) | -p2(B,A,C).
** KEPT (pick-wt=17): 12 [] -p2(A,B,C) |
       p2(B,f4(D,E,C,B,A),D) | -p2(D,E,C).
** KEPT (pick-wt=4): 13 [] -p2(c5,c6,c7).
** KEPT (pick-wt=36): 14 [] p1(A,B,C,D) | -p1(E,F,G,H) |
       -p2(E,F,A) | equal(E,F) | -p1(E,B,G,D) | -p2(G,H,C) |
-p1(F,B,H,D) | -p1(F,A,H,C).
** KEPT (pick-wt=20): 15 [] -p2(A,B,C) | equal(D,B) |
       -p2(D,B,E) | p2(D,C,f9(D,A,B,C,E)).

------------> process sos:
** KEPT (pick-wt=9):
       34 [] -p1(c12,c13,c12,f11(f11(c13,c12),c12)).
** KEPT (pick-wt=8): 35 [] p2(A,B,f3(A,B,C,D)).
** KEPT (pick-wt=9): 36 [] p1(A,f3(B,A,C,D),C,D).
** KEPT (pick-wt=9): 37 [] equal(f11(A,B),f3(A,B,A,B)).

---> New Demodulator:
       38 [new_demod,37] equal(f11(A,B),f3(A,B,A,B)).
** KEPT (pick-wt=5): 39 [] p1(A,B,B,A).
```

```
** KEPT (pick-wt=3): 40 [] equal(A,A).

>>>> Starting back demodulation with 38.

>> back demodulating 34 with 38.

39 back subsumes 33.

Following clause subsumed by 40 during input processing:
0 [copy,40,flip.1] equal(A,A).

======= end of input processing =======
=========== start of search ===========

-------- PROOF --------

----> UNIT CONFLICT at   0.53 sec
----> 154 [binary,153.1,145.1] $F.

Length of proof is 7.  Level of proof is 4.

---------------- PROOF ----------------

1 [] p2(A,f10(B,C,A,D,E,F),F) | -p1(B,E,B,F) |
       -p1(B,C,B,A) | -p2(C,D,E) | -p2(B,C,E).
2 [] -p1(A,B,C,D)|p1(E,F,C,D)| -p1(A,B,E,F).
3 [] -p2(A,B,C)|equal(D,B)| -p2(D,B,E) |
       p2(f8(D,A,B,C,E),E,f9(D,A,B,C,E)).
18 [factor,2,1,3] -p1(A,B,C,D) | p1(C,D,C,D).
34 [] -p1(c12,c13,c12,f11(f11(c13,c12),c12)).
36 [] p1(A,f3(B,A,C,D),C,D).
38,37 [] equal(f11(A,B),f3(A,B,A,B)).
39 [] p1(A,B,B,A).
41 [back_demod,34,demod,38,38]
       p1(c12,c13,c12,f3(f3(c13,c12,c13,c12),c12,f3
              (c13,c12,c13,c12),c12)).
49 [hyper,39,18] p1(A,B,A,B).
59 [hyper,36,2,49] p1(A,B,C,f3(D,C,A,B)).
60 [hyper,36,2,39] p1(A,B,f3(C,D,A,B),D).
145 [ur,41,2,59] -p1(f3(c13,c12,c13,c12),c12,c12,c13).
153 [hyper,60,2,39] p1(f3(A,B,C,D),B,D,C).
154 [binary,153.1,145.1] $F.

------------ end of proof -------------

Search stopped by max_proofs option.
============ end of search ============
```

```
-------------- statistics -------------
clauses given                  12
clauses generated            1557
clauses kept                  151
clauses forward subsumed     1419
clauses back subsumed          23
Kbytes malloced               287

----------- times (seconds) -----------
user CPU time          0.54          (0 hr, 0 min, 0 sec)
system CPU time        0.04          (0 hr, 0 min, 0 sec)
wall-clock time        0             (0 hr, 0 min, 0 sec)
hyper_res time         0.00
UR_res time            0.00
para_into time         0.00
para_from time         0.00
for_sub time           0.00
back_sub time          0.00
conflict time          0.00
demod time             0.00

The job finished        Wed Jul 16 14:06:58 1997

real          1.3          user          0.6       sys          0.0
```

## Exercises for Chapter 13

13.1. What predicates, functions, and constants do the symbols used in GEO026–2 correspond to?

13.2. What clauses constitute the set of "clauses in usable" for Gandalf? It might be easier to characterize the set by saying which clauses are not in the set.

13.3. In terms of the inferencing operations performed by THEO (that is, binary resolution and binary factoring), describe the inferences that THEO would carry out to obtain clause 88371 from clause 88350 and from clause 88349 as was done by Gandalf when proving GEO026–2.

13.4. When proving GEO041–2, Gandalf used the inferencing rule called paramodulation. This rule derived clause 49593 from clause 49583 and from clause 49540. THEO can also derive this clause from the same base clauses if additional clauses are used. What are these clauses? (Hint: the answer is related to the properties of the equality predicate.)

13.5. In proving GEO026–2, Otter derived a number of clauses on the way to finding a contradiction. These were clauses numbered: 20, 35, 38, 46, 74, 81, 84, 89, and 90. Show how THEO could have proved the theorem using its two inference rules and deriving the same sequence of intermediate inferences.

13.6. Attempt to prove GEO026–2, GEO041–2, and GEO059–2 using THEO. Compare the performance of THEO with Gandalf and Otter.

# Bibliography

Publications that cover material related to the contents of this book are listed below in order of their date of appearance.

N. Nilsson, *Problem-Solving Methods in Artificial Intelligence*, McGraw-Hill, New York, 1970.

C.L. Chang and R.C.T. Lee, *Symbolic Logic and Mechanical Theorem Proving*, Academic Press, New York, 1973.

D. Loveland, *Automated Theorem Proving*, North-Holland, Amsterdam, 1978.

J.A. Robinson, *Logic: Form and Function*, North-Holland,Amsterdam, 1979.

N. Nilsson, *Principles of Artificial Intelligence*, Tioga Press, Palo Alto, CA, 1980.

L. Wos et al., *Automated Reasoning: Introduction and Applications*, Prentice-Hall, Englewood Cliffs, NJ, 1984 (second Edition 1992 contains diskette with theorem-proving program Otter).

M. Genesereth and N. Nilsson, *Logical Foundations of Artificial Intelligence*, Morgan Kaufmann, San Francisco, 1987.

S. Tanimoto, *The Elements of Artificial Intelligence*, Computer Science Press, Rockville, MD, 1987.

S. Chou, *Mechanical Geometry Threorem Proving*, D. Reidel, Dordrecht, 1988.

L. Wos, *Automated Reasoning: 33 Basic Research Problems*, Prentice-Hall, Englewood Cliffs, NJ, 1988.

A number of journals publish material on automated theorem proving. First and foremost is the *Journal of Automated Reasoning*, published since 1985 by D. Reidel Publishing Company, P. O. Box 17, 3300 AA Dordrecht, Holland, or 190 Old Derby Street, Hingham, Massachusetts 02043.In addition, *Artifcial Intelligence*, published by North-Holland, *Machine Intelligence*, a series of almost a dozen volumes, which publishes every several years, the *Journal of the Association for Computing Machinery*, and the *IEEE Transactions on Computers* have all played leading roles in the publication of papers on automated theorem proving. The Conference on Automated Deduction is held once a year, and its proceedings are on the cutting edge of developments in the field.

Listed below are a number of other publications closely related to the material in this book.

J. Allen and D. Luckham, An interactive theorem-proving program, *Machine Intelligence* 5, (B. Meltzer and D. Michie, eds.), American Elsevier, New York, pp. 321–336, 1970.

R. Anderson and W.W. Bledsoe, A linear format for resolution with merging and a new technique for establishing completeness, *J. Assoc. Comput. Mach.* 17, 525–534, July 1970.

P. Andrews, Resolution with merging, *J. Assoc. Comput. Mach.* 15, 367–381, July 1968.

W.W. Bledsoe and P. Bruell, A man-machine theorem-proving system, *Artif. Intell.* 5, 51–72, 1974.

C.L. Chang, The unit proof and the input proof in theorem proving, *J. Assoc. Comput. Mach.* 17, 698–707, 1970.

S. Fleisig, A.K. Smiley III, and D.L. Yarmush, An implementation of the model elimination proof procedure, *J. Assoc. Comput. Mach.* 21, 124–139, 1974.

S. Greenbaum, D.A. Plaisted, and J.H. Siekmann, The Illinois Prover: a general purpose resolution theorem prover, *Proceedings of the 8th International Conference on Automated Deduction*, Oxford, Springer-Verlag, New York, pp. 685–686, 1986.

R.A. Kowalski and D. Kuehner, Linear resolution with selection function, *Artif. Intell.* 2, 147–178, 1971.

J.D. Lawrence and J.D. Starkey, Experimental tests of resolution-based theorem proving strategies, CS-74-011, Computer Science Department, Washington State University, April 1974.

R. Letz, J. Schumann, S. Bayerl, and W. Bibel, SETHEO, a high performance theorem prover, *J. Autom. Reasoning* 8, 183–213, 1992.

D.W. Loveland, A linear format for resolution, Symposium on Automatic Deduction, *Lecture Notes in Mathematics*, 125, Springer-Verlag, Berlin, 1968.

D.W. Loveland, Automated theorem proving: a quarter century review, *Contemporary Mathematics* 29, 1984 (presented at the 1984 American Mathematical Society's Special Session on Automated Theorem Proving, Denver, CO, 1983).

D. Luckham, Some tree-paring strategies for theorem proving, *Machine Intelligence* 3, (D. Michie, ed.) American Elsevier, New York, pp. 95–112, 1968.

J.D. McCharen, R.A. Overbeek, and L.A. Wos, Problems and experiments for and with automated theorem-proving programs, *IEEE Trans. Comput.* 25, 773–783, 1976 (contains many theorems used for testing theorem-proving programs).

H.J. Ohlbach and M. Schmidt-Schauss, The lion and the unicorn (theorem proving), *J. Autom. Reasoning* 1, 327–332, 1985.

H.J. Ohlbach, Predicate logic hacker tricks, *J. Autom. Reasoning* 1, 435–440, 1985.

M. Paterson and M. Wegman, Linear unification, *J. Comput. Syst. Sci.* 16, 158-167, 1968.

F.J. Pelletier, Seventy-five problems for testing automatic theorem provers, *J. Autom. Reasoning* 2, 191–216, 1986.

A. Quaife, Automated development of Tarski's geometry, *J. Autom. Reasoning* 5, 97–118, 1989.

J. Robinson, A machine-oriented logic based on the resolution principle, *J. Assoc. Comput. Mach.* 12, 23–41, 1965.

J. Siekmann and G. Wrightson(eds.), *The Automation of Reasoning*, Vols. 1 and 2, Springer-Verlag, New York, 1983.

R. Socher, Optimizing the clausal normal form transformation, *J. Autom. Reasoning* 7, 325–336, 1991.

M. Stickel, A Prolog Technology Theorem Prover: implementation by an extended Prolog compiler, *J. Autom. Reasoning* 4, 353–380, 1988.

M. Stickel, Schubert's steamroller problems: formulations and solutions, *J. Autom. Reasoning* 2, 89–101, 1986.

L. Wos, D. Carson, and G.A. Robinson, The unit preference strategy in theorem proving, *AFIPS Conference Proceedings 26*, Spartan Books, Washington, DC, pp. 615–621, 1964.

L. Wos, J.A. Robinson, and D.F. Carson, Efficiency and completeness of the set-of-support strategy in theorem proving, *J. Assoc. Comput. Mach.* 12, 536–541, October 1965.

L. Wos et al., An overview of automated reasoning and related fields, *J. Autom. Reasoning* 1, 5–48, 1985.

L. Wos, Automated reasoning answers open questions, *Not. Am. Math. Soc.* 40, 15–26, January 1993.

# Appendix A
# Answers to Selected Exercises

## Chapter 1

1.2. This problem and the next are meant to familiarize you with the TPTP Problem Library. There are 28 categories of theorems. See Documents/OverallSynopsis in the TPTP Problem Library.

1.3. There are 160 geometry theorems in the directory GEO.

## Chapter 2

2.1. (a)
```
@x: {positiveinteger(x) & perfectsquare(x)} =>
        !y: {equal(times(plus(y,1),minus(y,1)),plus(x,1))
```
(b)
```
@x: {{positiveinteger(x) & ~prime(x)} =>
        !y: {prime(y) & divides(y,x) & lessthan(y,x)}}
```
(c)
```
on(c,b) | {!x: on(c,x) & above(x,b)}  =>  above(c,b)
```

2.2. See HUMBIRD.WFF.

2.3. See MERMAID.WFF.

2.4. See NATNUMB.WFF.

2.5. There are a total of 21 axioms.

2.6. See Q49C2.THM.

2.7. See Q59W3.THM.

2.8. See CANNIBAL.WFF.

2.9. See PUZZLE8.WFF.

## Chapter 3

3.1. (a)
```
~E(x,y) | E(y,x)
```
(b)
```
~A(x) | ~B(x) | ~C(x) | D(x)
```

(c) ~A(x) | B(x)
~A(x) | C(x)
~A(x) | D(x)

(d) ~A(x) | B(x) | C(x) | D(x)

(e) ~A(x) | D(x)
~B(x) | D(x)
~C(x) | D(x)

(f) - (m) use COMPILE

3.2. Run COMPILE on NATNUMB.THM.
3.3. See S25WOS1.THM.
3.4. See Q01D1.THM.

## Chapter 4

4.1. (c) {f(z)/u,f(z)/x}, Q(f(z),f(z))
(d) {a/z,f(a)/x,g(y)/u}, P(a,f(a),f(g(y)))
(e) {f(a)/x,f(f(a))/y,f(f(f(a)))/z,f(f(f(f(a))))/v,f(f(f(f(f(a)))))/u},
R(f(a),f(f(a)),f(f(f(a))),f(f(f(f(a)))),f(f(f(f(f(a))))))}

4.2. (c) C: (5c,6b) ~less(a,u) | ~less(f(a),u) | less(h(u),u)
(d) C: (7b,8a) P(x,x) | ~Q(x,y) | Q(f(x),a)
C: (7c,8b) P(f(u),f(u)) | ~P(f(u),h(f(u))) | P(u,v)
(e) C: (9a,10a) Q(y,h(y)) | Q(g(u,a),b) | P(u) | ~Q(b,h(a))

4.3. (b) C: (12cd) P(a,f(a)) | P(f(a),f(b)) | Q(f(a),a)
(c) C: (13ac) P(a,f(a)) | P(f(a),f(a))
C: (13bc) P(a,a) | P(a,f(a))

4.4. k*m. Consider C1 = P(a,x) | P(b,x) | P(c,x) and C2 = ~P(x,a) | ~P(x,b) | ~P(x,c).

4.5. k*(k - 1)/2. Consider C = P(x,y,z) | P(x,y,a) | P(x,b,a) | P(c,b,a).

4.7. (a) subsumes and s-subsumes, (b) subsumes and s-subsumes,
(c) subsumes, (d) does not subsume,
(e) does not subsume, (f) does not subsume,
(g) does not subsume, (h) subsumes,
(i) does not subsume, (j) does not subsume.

4.9. Consider C1 = ~P(x) | P(f(x)) and C2 = ~P(x) | P(f(f(x)))

## Chapter 5

5.1. Herbrand universe elements:1 a 2 h(a,a) 3 h(a,h(a,a)) 4 h(h(a,a),a)
5 h(h(a,a),h(a,a)) 6 h(a,h(a,h(a,a))) 7 h(a,h(h(a,a),a)) 8 h(a,h((a,a)h(a,a)))
9 h(h(a,a),h(a,h(a,a))) 10 h(h(a,a),h(h(a,a),a)) 11 h(h(h(a,a),a),h(h(a,a),a))
12 h(h(a,h(a,a),a)) 13 h(h(a,h(a,a),h(a,a)) 14 h(h(a,h(a,a)),h(a,h(a,a)))
15 h(h(a,h(a,a)),h(h(a,a),a)),16 h(h(a,h(a,a)),h(h(a,a),h(a,a)))
17 h(h(h(a,a),a),a) 18 h(h(h(a,a),a),h(a,a)) 19 h(h(h(a,a),a),h(a,h(a,a)))
20 h(h(h(a,a),a),h(h(a,a),a))
Herbrand base elements: 1 P(a,a) 2 Q(a,a) 3 P(a,h(a,a)) 4 Q(a,h(a,a))
5 P(h(a,a),a) 6 Q(h(a,a),a) 7 P(a,h(a,h(a,a))) 8 Q(a,h(a,h(a,a))) 9 P(h(a,a),h(a,a))
10 Q(h(a,a),h(a,a)) 11 P(h(a,h(a,a)),a) 12 Q(h(a,h(a,a)),a) 13 P(a,h(h(a,a),a))
14 Q(a,h(h(a,a),a)) 15 P(h(a,a),h(a,h(a,a))) 16 Q(h(a,a),h(a,h(a,a)))
17 P(h(a,h(a,a)),h(a,a)) 18 Q(h(a,h(a,a)),h(a,a)) 19 P(h(h(a,a),a),a)
20 P(h(h(a,a),a),a)

5.2. There are two terms in the Herbrand universe: HU = {a,e}. There are 31 atoms in the Herbrand base.

5.4.

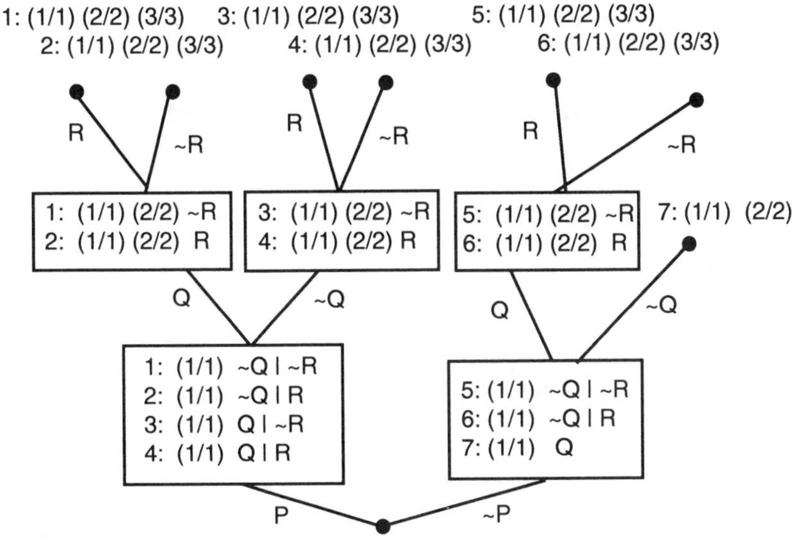


5.5. Use atom P(a).

5.6. Use atoms P(a), P(g(a)), P(g(g(a))), and P(g(g(g(a)))).
5.7. Use atoms Equal(e,a), P(e,a,a), and P(e,a,e).

## Chapter 6

6.1. @x: ~Prime(x) => {!y: Divides(y,x) & Prime(y) & Less(y,x)}

6.2. Step 1: Form resolvent C8: (1c,2c) ~P | ~Q. This yields the semantic tree:

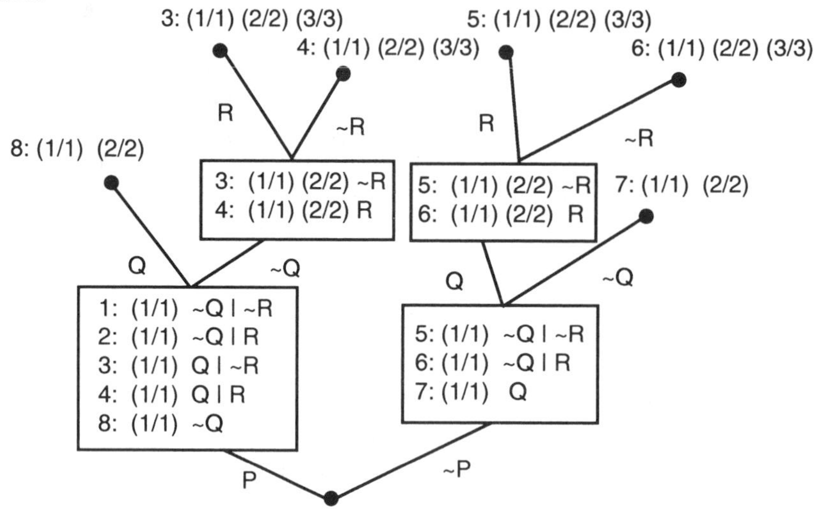


Step 2: Form resolvent C9: (3c,4c) ~P | Q. This gives the semantic tree:

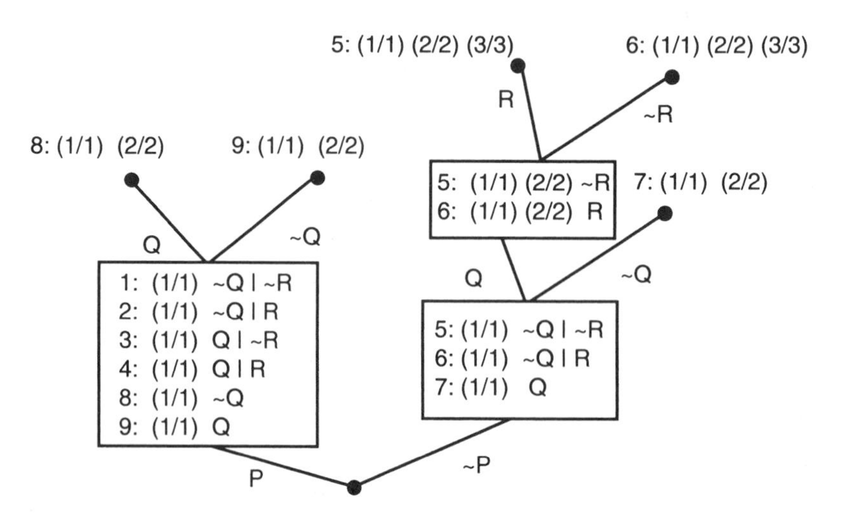

Step 3: Form resolvent C10: (5c,6c) P | ~Q. This gives the semantic tree:

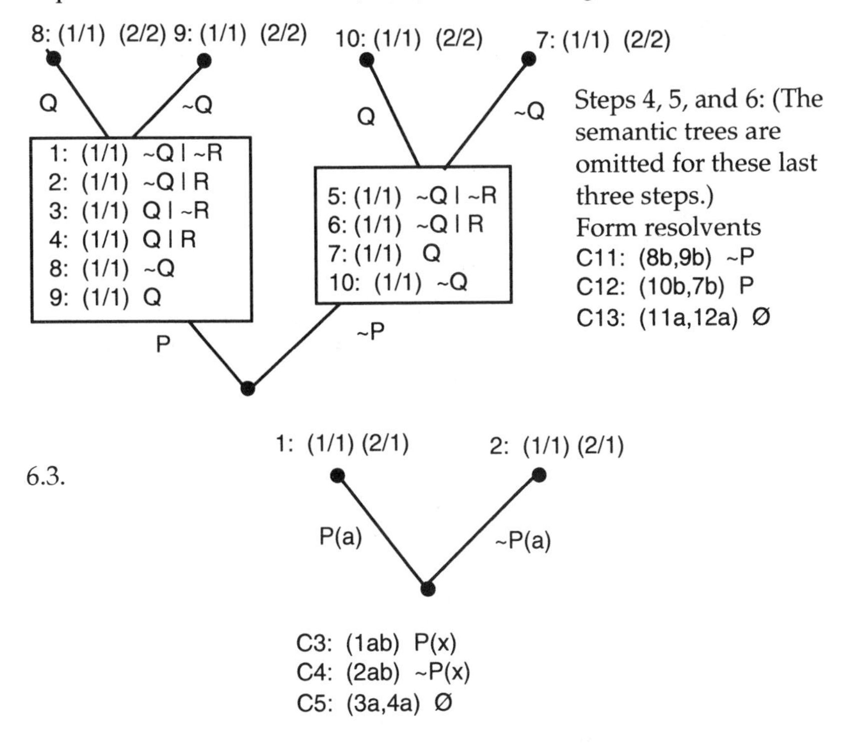


6.4.

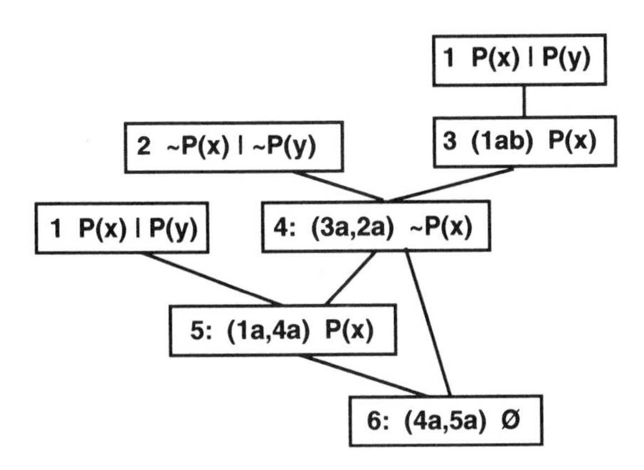


6.5. This is a difficult problem, but lots of fun!

6.6. $2^{D+1} - 2 - D$

6.11.

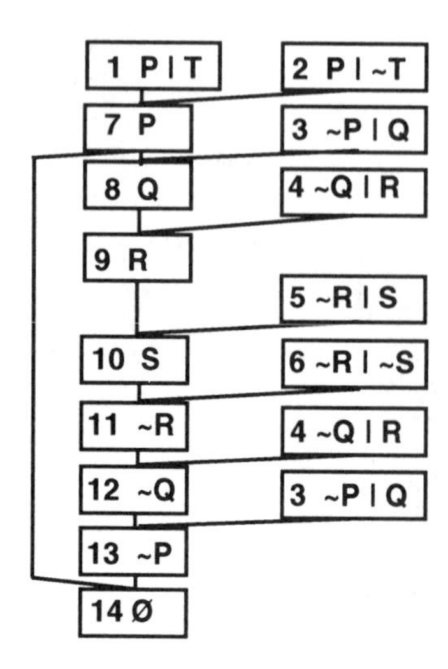


## Chapter 7

7.2. Examine the source code in atom.c.

7.3. Run HERBY on STARK036.THM using the d2 option. Whenever the execution of the program halts, enter a carriage return. From the printout, you can then construct the tree by hand (36 nodes in the tree) and show where the tree-pruning heuristic of Section 7.2.5 comes into play.

## Chapter 8

8.1. The heuristic was used.

8.4. Examine the proof found by THEO and see if that gives you any ideas for selecting atoms.

## Chapter 9

9.1. Check your results by running THEO with the options p20 k0 h0 n0 m0 z0. The results can be found in the file B.T0. (Note: if your computer is slow, the results may be placed in B.T1, where the 1 indicates your computer took one second to prove the theorem; a 0 indicates that less than one second was necessary.) Two iterations are necessary, and seven nodes are generated.

9.2. Check your results by running THEO with the options p20 k0 h0 n0 z1. THEO carries out three iterations and generates seven clauses when proving B.THM. THEO carries out five iterations and generates 18 clauses when proving C.THM.

9.3. Check your results by running THEO with the options p20 k0 h0 z1 b0. THEO carries out one iterations and generates three clauses when proving B.THM. THEO carries out three iterations and generates 11 clauses when proving C.THM.

9.4. P(x,f(a)) => 15, ~Q(y,f(x)) => –19 ~P(x,y) => –103, ~Q(z,y)=> –106, Q(a,x).

9.5. For LITmaxbase = 2, the following is a counterexample: 1 p | s, 2 ~s | r, 3 ~p | r, 4 ~p | t, 5 ~t | ~r, 6 ~p | ~r. Now, try LITmaxbase = 3 for yourself!

9.6. There are four constants, e,a, b, and c. Clauses 1, 2, 6,16, and 17 contain one variable; for each there are five hash codes entered in the clause_hash_table. For clauses 3 which has two variables, there are 25 hash codes entered, and for clauses 20, 21, 22, and 23 there is one each.

## Chapter 10

10.1. (b) Sixteen entries are made in clause_hash_table. (c) The two unit base clauses, UNAAB and EAB, account for two, then during the search, entries were made for clauses ~EAB (1 entry), Sxx (4 entries corresponding to Sxx, Syy, SAA, and SBB), Exx (4 entries), SBA (1 entry), ~SAB (1 entry), MFABA (coming from the clause SAx | MFAxA after literal SAx was resolved away as discussed in Section 9.14), MFABB (1 entry), and SAB (1 entry) for a total of 16. (Note: this problem is impossible to answer completely without understanding some of the source couse in the file search.c.)

10.7. The four theorems are in the diretory THEOREMS. If you create PROVEALL1 as shown here, all the theorems can be attempted without user interaction. An upper limit of 1220 seconds is placed on each attempt to find a proof.

```
STARK075.THM z0 h0 n0 t1200
STARK075.THM z0 h0 n1 t1200
STARK075.THM z0 h1 n0 t1200
```

```
STARK075.THM z0 h1 n1 t1200
STARK075.THM z1 h0 n0 t1200
STARK075.THM z1 h0 n1 t1200
STARK075.THM z1 h1 n0 t1200
STARK075.THM z1 h1 n1 t1200
   .
   .
   .
S46WOS22.THM z1 h1 n1 t1200
```

10.8. The theorem can be found in the directory THMSMISC.

10.9. First, convert the wffs cannibal.wff and puzzle8.wff in the directory THMSMISC to clauses and then use THEO to obtain proofs.

10.10, 10.11. The theorems can be found in the directory THMSMISC.

## Chapter 13

13.1. These are clauses that are not equality substitution clauses.

13.2. The first two literals of clause 88350 are factored and then the result resolved with clause 88349.

13.3. The clauses ~Equal(x,y) | Equal(y,x) and ~Equal(x,y) | ~p2(u,v,x) | p2(u,v,y).

# Appendix B
# List of Wffs and Theorems in the Directories WFFS, THEOREMS, GEOMETRY, and THMSMISC

The directory WFFS contains 21 files, each containing at least one wff:

1. CANNIBAL.WFF ;Exercises 2.8, 10.9: missionaries and cannibals
2. EXCOMP1.WFF ; Chapter 3.1: running example
3. EXCOMP2.WFF ; Example 1, Section 3.3
4. EXCOMP3.WFF ; Example 2, Section 3.3
5. EXCOMP4.WFF ; Example 3, Section 3.3
6. EXCOMP5.WFF ; Example 4, Section 3.3
7. EXER3F.WFF ; Exercise 3.1f
8. EXER3G.WFF ; Exercise 3.1g
9. EXER3H.WFF ; Exercise 3.1h
10. EXER3I.WFF ; Exercise 3.1i
11. EXER3J.WFF ; Exercise 3.1j
12. EXER3K.WFF ; Exercise 3.1k
13. EXER3L.WFF ; Exercise 3.1l
14. EXER3M.WFF ; Exercise 3.1m
15. EXER3N.WFF ; Exercise 3.1n
16. EXER3O.WFF ; Exercise 3.1o
17. HUMBIRD.WFF ; Exercise 2.2
18. MERMAID.WFF ; Exercise 2.3
19. NATNUMB.WFF ; Exercise 2.4, 3.2
20. PUZZLE8.WFF ; Exercises 2.9, 10.9: the 8-puzzle
21. WOLVES.WFF ; Exercise 10.12

The directory THEOREMS contains 85 files, 84 each containing one theorem of the Stickel test set. One file is intended for use in proving the Stickel set without user intervention. These theorems have been used by several researchers when testing their theorem-proving programs.

1. S01BURST.THM
2. S02SHORT.THM
3. S03PRIM1.THM
4. S04HASP1.THM
5. S05HASP2.THM
6. S06ANCES.THM
7. S07NUM1.THM
8. S08GRP1.THM
9. S09GRP2.THM
10. S10EW1.THM
11. S11EW2.THM
12. S12EW3.THM
13. S13ROB1.THM
14. S14ROB2.THM
15. S15MICHI.THM
16. S16QW.THM
17. S17MQW.THM
18. S18DBABH.THM
19. S19APABH.THM
20. S20FLEI1.THM
21. S21FLEI2.THM
22. S22FLEI3.THM
23. S23FLEI4.THM
24. S24FLEI5.THM
25. S25WOS1.THM
26. S26WOS2.THM
27. S27WOS3.THM
28. S28WOS4.THM
29. S29WOS5.THM
30. S30WOS6.THM
31. S31WOS7.THM
32. S32WOS8.THM
33. S33WOS9.THM
34. S34WOS10.THM
35. S35WOS11.THM
36. S36WOS12.THM
37. S37WOS13.THM
38. S38WOS14.THM
39. S39WOS15.THM
40. S40WOS16.THM
41. S41WOS17.THM
42. S42WOS18.THM
43. S43WOS19.THM
44. S44WOS20.THM
45. S45WOS21.THM
46. S46WOS22.THM
47. S47WOS23.THM
48. S48WOS24.THM
49. S49WOS25.THM
50. S50WOS26.THM
51. S51WOS27.THM
52. S52WOS28.THM
53. S53WOS29.THM
54. S54WOS30.THM
55. S55WOS31.THM
56. S56WOS32.THM
57. S57WOS33.THM

| | |
|---|---|
| 58. STARK005.THM | ; Lawrence and Starkey #5, Nilsson (1970), p. 220. |
| 59. STARK017.THM | ; There exist infinitely many primes. |
| 60. STARK023.THM | ; In a group with right inverses and right iden-<br>; tity, every element has a left inverse. |
| 61. STARK026.THM | ; |
| 62. STARK028.THM | ; (a + b) + c = a + (b + c) |
| 63. STARK029.THM | ; (a - b) + c =(a + c) - b |
| 64. STARK035.THM | ; |
| 65. STARK036.THM | ; In a group,(x times y) inverse equals x inverse<br>; times y inverse. |
| 66. STARK037.THM | ; In a ring, x*0 = 0. |
| 67. STARK041.THM | ; Equality is symmetric. |
| 68. STARK055.THM | ; Kleene, Intro. to Metamathematics. |
| 69. STARK065.THM | ; If x is less than y and y is less than or equal to<br>; z, then x is less than z. |
| 70. STARK068.THM | ; 0 < x'. |
| 71. STARK075.THM | ; |
| 72. STARK076.THM | ; If x < y, then y < x. |
| 73. STARK087.THM | ; If z≠ 0, then if xz < yz, x < y. |
| 74. STARK100.THM | ; If x = y, then every member of x is also a<br>; member of y. |
| 75. STARK103.THM | ; Union is idempotent. |
| 76. STARK105.THM | ; If (x union y) = y, then x is a subset of y. |
| 77. STARK106.THM | ; x is the union of x and y. |

78. STARK108.THM ; Intersection is associative.
79. STARK111.THM ; If the intersection of x and y is x, then x is a ; subset of y.
80. STARK112.THM ; x union (y intersection z) equals (x intersec- ; tion y) union (x intersection z).
81. STARK115.THM ; Intersection of x & (y – x) is empty.
82. STARK116.THM ; If x is in y, then z – y is in z – x.
83. STARK118.THM ; (z – x) union (z – y) = z – (x intersection y).
84. STARK121.THM ; x intersection y equals x – (x – y).

The directory GEOMETRY contains 66 files, each containing one theorem. These theorems on Euclidean geometry are the subject of Art Quaife's paper, "Automated development of Tarski's geometry" (listed in the bibliography). These theorems are based on the axioms presented in Section 2.6. Each successive theorem uses all the previous ones as axioms.

1. Q01D1.THM ; ordinary reflexivity of equidistance
2. Q02D2.THM ; equidistance is symmetric between argument pairs
3. Q03D3.THM ; equidistance is symmetric with argument pairs
4. Q04D4.THM ; corollaries to Q02D2.THM and Q03D3.THM
5. Q05D5.THM ; ordinary transitivity of equidistance
6. Q06E1.THM ; NULL extension
7. Q07B0.THM ; corollary to axiom A4.1
8. Q08R2.THM ; corollaries to axioms A4
9. Q09R3.THM ; corollaries to Q06E1.THM
10. Q10R4.THM ; u is the only fixed point of R(u,v)
11. Q11D7.THM ; all NULL segments are congruent
12. Q12D8.THM ; addition of equal segments
13. Q13D9.THM ; unique extension
14. Q14D10A.THM ; corollaries to Q13D9.THM
15. Q14D10B.THM ; " "
16. Q14D10C.THM ; " "
17. Q15R5.THM ; congruence for double reflection
18. Q16R6.THM ; R is an involution
19. Q17T3.THM ; v is between u and v
20. Q18B1.THM ; corollary with axiom A6
21. Q19T1.THM ; between is symmetric in its outer arguments
22. Q20T2.THM ; u is between u and v
23. Q21B2.THM ; antisymmetry of between in first two arguments
24. Q22B3.THM ; corollary to Q21B2.THM

25. Q23T6.THM ; given three distinct points u, v, w, if v is
 ; between u and w, then u is not between v and
 ; w, and w is not between u and v.
26. Q24B4.THM ; first inner transitivity property of betweenness
27. Q25B5.THM ; corollary using symmetry
28. Q26B6.THM ; first outer transitivity property of betweenness
29. Q27B7.THM ; second outer transitivity property of betweenness
30. Q28B8.THM ; second inner transitivity property of betweenness
31. Q29B9.THM ; corollary using symmetry
32. Q30E2.THM ; there are at least three distinct points
33. Q31E3.THM ; a segment can be lengthened
34. Q32B10.THM ; inner points of a triangle
35. Q33D11.THM ; corollary to outer five-segment axiom
36. Q34D12.THM ; second inner five-segment theorem
37. Q35D13.THM ; subtraction of line segments
38. Q36D14.THM ; first inner five-segment theorem
39. Q37D15.THM ; corollary
40. Q38I2A.THM ; theorem of point insertion
41. Q38I2B.THM ; " "
42. Q38I2C.THM ; " "
43. Q39I3.THM ; insertion identity
44. Q40I4.THM ; insert respects congruence in its last two arguments
45. Q41B11.THM ; theorem of similar situations
46. Q42B12.THM ; first outer connectivity property of betweenness
47. Q43B13.THM ; second outer connectivity property of betweeneess
48. Q44T7.THM ; given two distinct points w, x, and two distinct
 ; points u, v on the segment wx, to the left of w,
 ; then either v is between u and w, or u is
 ; between v and w.
49. Q45T9.THM ; first inner connectivity property of betweenness
50. Q46B14.THM ; second inner connectivity property of betweenness
51. Q47T8.THM ; five-point theorem
52. Q48B15.THM ; unique endpoint
53. Q49C2.THM ; collinearity: corollary using symmetry
54. Q50T10.THM ; collinearity is invariant under permutation of
 ; arguments
55. Q51T11.THM ; the three points p', p", and p"' are not collinear
56. Q52C3.THM ; any two points are collinear
57. Q53C4.THM ; theorem of similar situation for C(u,v,w)
58. Q54T12.THM ; given three distinct collinear points u,v,w, and
 ; a fourth point x collinear with u and v, then x is
 ; collinear with uw and collinear with vw.

59. Q55C5.THM ; additional versions using symmetry
60. Q56T13.THM ; given two distinct points u,v and three points
; w, w', w" which are collinear with uv, then
; w, w', w" are collinear.
61. Q57W1A.THM ; lemma: under the hypothesis of the bisecting
; diagonal theorem, the points u,v,w cannot be
; collinear
62. Q57W1B.THM ; " "
63. Q57W1C.THM ; " "
64. Q58W2A.THM ; corollaries
65. Q58W2B.THM ; " "
66. Q59W3.THM ; the diagonals of a nondegenerate rectangle
; bisect each other

The directory THMSMISC contains 32 files, each containing one theorem. These are miscellaneous theorems that appear in the text or in various research papers.

1. CHANG1.THM
2. CHANG2.THM
3. CHANG3.THM
4. CHANG4.THM
5. CHANG5.THM
6. CHANG6.THM
7. CHANG7.THM
8. CHANG8.THM
9. CHANG9.THM
10. AGATHA.THM
11. LION.THM
12. KNGTS.THM
13. WOLVES1.THM
14. WOLVES2.THM
15. WOLVES3.THM
16. SAM.THM
17. LOVE39A.THM
18. LOVE39B.THM
19. LOVE39C.THM
20. LIFSCHTZ.THM
21. EQUAL1.THM
22. EQUAL2.THM
23. A.THM
24. B.THM
25. C.THM
26. FACT.THM
27. SQROOT.THM
28. HU1.THM
29. HU2.THM
30. LINEAR.THM
31. HASH.THM
32. EX2.THM

# Index